Swing Extreme Testing

The Extreme Approach to Complete Java Application Testing

Tim Lavers

Lindsay Peters

BIRMINGHAM - MUMBAI

Swing Extreme Testing

Copyright © 2008 Packt Publishing

All rights reserved. No part of this book may be reproduced, stored in a retrieval system, or transmitted in any form or by any means, without the prior written permission of the publisher, except in the case of brief quotations embedded in critical articles or reviews.

Every effort has been made in the preparation of this book to ensure the accuracy of the information presented. However, the information contained in this book is sold without warranty, either express or implied. Neither the author(s), Packt Publishing, nor its dealers or distributors will be held liable for any damages caused or alleged to be caused directly or indirectly by this book.

Packt Publishing has endeavored to provide trademark information about all the companies and products mentioned in this book by the appropriate use of capitals. However, Packt Publishing cannot guarantee the accuracy of this information.

First published: May 2008

Production Reference: 1260508

Published by Packt Publishing Ltd.
32 Lincoln Road
Olton
Birmingham, B27 6PA, UK.

ISBN 978-1-847194-82-4

www.packtpub.com

Cover Image by Vinayak Chittar (vinayak.chittar@gmail.com)

Credits

Authors
Tim Lavers
Lindsay Peters

Reviewers
Prabhakar Chaganti
Valentin Crettaz

Senior Acquisition Editor
David Barnes

Development Editor
Ved Prakash Jha

Technical Editor
Himanshu Panchal

Editorial Team Leader
Mithil Kulkarni

Project Manager
Abhijeet Deobhakta

Project Coordinator
Lata Basantani

Indexer
Hemangini Bari

Proofreader
Angie Butcher

Production Coordinator
Aparna Bhagat

Cover Work
Aparna Bhagat

About the Authors

Tim Lavers is a Senior Software Engineer at Pacific Knowledge Systems, which produces LabWizard, the gold-standard for rules-based knowledge acquisition software. In developing and maintaining LabWizard for almost ten years, Tim has worked with many Java technologies including network programming, Swing, reflection, logging, JavaHelp, Web Services, RMI, WebStart, preferences, internationalization, concurrent programming, XML, and databases as well as tools such as Ant and CruiseControl. His job also includes a healthy mix of user training, technical support, and support to marketing. In his previous job, he wrote servlets and built an image processing library. Along with his professional programming, he writes and maintains the distributed testing tool 'GrandTestAuto'. Previously he has published a JavaWorld article on RMI as well as a number of mathematical papers.

Tim's hobbies include running and playing the piano.

Lindsay Peters is the Chief Technical Officer for Pacific Knowledge Systems. He has 25 years experience in software management, formal analysis, algorithm development, software design, and implementation for large commercial and defense systems. Ten years ago Lindsay and his team were early adopters of Java, coupled with more rigorous design processes such as Design by Contract. He then helped transition the development team to the Extreme Programming model. Out of this exciting and successful experience grew the "Extreme Testing" approach.

In the early 80's Lindsay managed a software team who were one of the first to incorporate the newly discovered simulated annealing algorithm into a commercial application, solving a previously intractable real-world problem which was the optimum assignment of radio frequencies to collocated mobile radios.

Apart from software development and artificial intelligence systems, Lindsay has an interest in mathematical convexity and has helped to progress the "Happy Ending" problem. He is also involved in politics and in the last Australian Federal election stood as the Greens candidate for the seat of Bennelong.

We gratefully acknowledge the support of Pacific Knowledge Systems in providing a software development environment that fostered the development of the Java Extreme Testing approach. We are also grateful to Pacific Knowledge Systems for allowing us to use code examples and screen shots from the LabWizard product.

We would like to thank the technical reviewers for their many excellent suggestions. Any remaining shortcomings are, of course, our own.

Thanks also to the team at Packt: David, Ved, Lata, and Himanshu. Their patience and responsiveness have made our job a lot easier and more enjoyable than it might otherwise have been.

Tim would like to thank his wife, Lorien, for her continual support and encouragement with this project. As well as a morale-building role, she has helped with the wording of several passages in the book.

Finally, Lindsay would like to thank his wife Kath for cheerfully picking up all the slack, and his father Harry for his continued inspiration.

To our children

Isobel, Hugh and Julia

and

Francis, Mariel, Vicente, Amparo, Bridie and Harry

About the Reviewers

Prabhakar Chaganti is the CTO of HelixBrain—a unique startup that provides technology services consulting and is an incubator nurturing some very cool software as service applications that are being built on the Ruby on Rails platform. His interests include Linux, Ruby, Python, Java, and Virtualization. He won the community choice award for the most innovative virtual appliance in the 2006 VMWare Ultimate Global Virtual Appliance Challenge. He is also the author of "GWT Java AJAX Programming" published in 2007.

Valentin Crettaz holds a master degree in Information and Computer Science from the Swiss Federal Institute of Technology in Lausanne, Switzerland (EPFL). After he finished studying in 2000, Valentin worked as a software engineer with SRI International (Menlo Park, USA) and as a principal engineer in the Software Engineering Laboratory at EPFL. In 2002, as a good patriot, he came back to Switzerland to co-found a start-up called Condris Technologies, a company that provides IT development and consulting services and specializes in the creation of innovative next-generation software architecture solutions as well as secure wireless telecommunication infrastructures.

Since 2004, Valentin serves as a senior IT consultant in one of the largest private banks in Switzerland, where he works on next-generation e-banking platforms. Starting in 2008, Valentin created Consulthys, a new venture that strongly focuses on leveraging Web 2.0 technologies in order to reduce the cultural gap between IT and business people.

Valentin's main research interests include service-oriented architecture and web services, Web 2.0, and Ajax. During his spare time, he hangs out as a moderator at JavaRanch.com and JavaBlackBelt.com, two of the largest Java development communities on the web. Valentin also serves as a technical editor at Manning Publications (CT, US) and as a technical reviewer at O'Reilly & Associates (CA, US) and Packt Publishing (UK).

Table of Contents

Preface	**1**
Chapter 1: What Needs Testing?	**13**
An Example	14
What Classes Do We Test?	16
Test First—Always!	19
What Classes Don't We Test?	21
What Methods Need Testing?	22
What Methods Don't We Test?	24
Invoking Hidden Methods and Constructors	25
Unit Test Coverage	27
Who Should Implement the Unit Tests?	28
What About Legacy Code?	29
Where Does Integration Testing Fit In?	29
Documentation of Unit Tests	29
Testing at the Application Level	30
Who Should Implement the Function Tests?	31
Automated Test Execution	31
A Hierarchy of Tests	33
What Language Should Our Tests Be In?	34
Is it Really Possible?	34
Summary	36
Chapter 2: Basics of Unit Testing	**37**
A Simple Example	37
The Default Implementation	38
Test Cases	39
Design by Contract and Non-Defensive Programming	40
Test Code Example	45
Bootstrapping Our Implementation	48

Load Testing	49
Summary	49

Chapter 3: Infrastructure for Testing — 51

Where Should the Unit Tests Go?	51
Where Should the Function and Load Tests Go?	53
Management of Test Data	54
What Do We Require of a Test Data Management System?	55
Temporary Files	57
Summary	58

Chapter 4: Cyborg—a Better Robot — 59

The Design of Cyborg	59
Using the Keyboard	60
Mousing Around	63
Checking the Screen	65
Summary	66

Chapter 5: Managing and Testing User Messages — 67

Some Problems with Resource Bundles	67
A Solution	69
The UserStrings Class	70
ResourcesTester	73
How ResourcesTester Works	74
Getting More from UserStrings	78
Summary	79

Chapter 6: Making Classes Testable with Interfaces — 81

The LabWizard Comment Editor	81
The Wizard	83
A Test for Wizard	85
A Test for Step	86
Handlers in LabWizard	90
Summary	90

Chapter 7: Exercising UI Components in Tests — 91

The LabWizard Login Screen	91
The Design of LoginScreen	94
UI Wrappers	96
The Correct Implementation of UILoginScreen	98
A Handler Implementation for Unit Testing	99
Setting Up our Tests	100
Our First Test	102
Further Tests	104

Some Implicit Tests ... 105
Other User Interfaces ... 105
Summary ... 105

Chapter 8: Showing, Finding, and Reading Swing Components — 107
Setting Up User Interface Components in a Thread-Safe Manner ... 108
Finding a Component ... 110
Testing Whether a Message is Showing ... 112
Searching for Components by Name ... 113
Reading the State of a Component ... 114
Case Study: Testing Whether an Action Can Be Cancelled ... 115
The Official Word on Swing Threading ... 117
Summary ... 118

Chapter 9: Case Study: Testing a 'Save as' Dialog — 119
The Ikon Do It 'Save as' Dialog ... 119
Outline of the Unit Test ... 121
UI Helper Methods ... 122
  Dialogs ... 123
  Getting the Text of a Text Field ... 124
  Frame Disposal ... 125
Unit Test Infrastructure ... 125
  The UISaveAsDialog Class ... 125
  The ShowerThread Class ... 127
  The init() Method ... 128
  The cleanup() Method ... 129
The Unit Tests ... 129
  The Constructor Test ... 130
  The wasCancelled() Test ... 131
  The name() Test ... 133
  The show() Test ... 134
  The Data Validation Test ... 135
  The Usability Test ... 136
  Summary ... 137

Chapter 10: More Techniques for Testing Swing Components — 139
Testing with JColorChooser ... 139
Using JFileChooser ... 142
Checking that a JFileChooser has been Set Up Correctly ... 142
Testing the Appearance of a JComponent ... 144
Testing with Frames ... 147
  Frame Location ... 147
  Frame Size ... 149

Testing with Lists	**150**
List Selection Methods	150
List Rendering	151
List Properties	153
Testing a JTable	**153**
Testing with JMenus	**156**
Checking the Items	156
Using Menus with Cyborg	159
Testing JPopupMenus	**160**
Combo Boxes	**160**
Progress Bars	**161**
JSlider and JSpinner	**163**
JTree	**164**
Summary	**166**

Chapter 11: Help!   167

Overview	**168**
What Tests Do We Need?	**169**
An HTML File That is Not Indexed	170
An index item for which there is no HTML file	171
Broken links	171
Incorrectly titled help pages	171
Creating and Testing Context-Sensitive Help	**172**
Executing HelpGenerator	**175**
Summary	**176**

Chapter 12: Threads   177

The Waiting Class	**177**
Concurrent Modifiers	**179**
Concurrent Readers and Writers	**181**
Proof of Thread Completion	**183**
The Unit Test for waitForNamedThreadToFinish()	**187**
Counting Threads	**189**
Summary	**190**
Further Reading	**190**

Chapter 13: Logging   191

Logging to a File	**191**
Remember to Roll!	**193**
Testing What is Printed to the Console	**193**
Switching Streams	194
Reading the Output From a Second JVM	196
Summary	**199**

Chapter 14: Communication with External Systems — 201
- Email — 201
  - Using an External Email Account — 202
  - Using a Local Email Server — 205
  - Which Method is Best? — 206
- Testing Spreadsheets — 206
- PDF — 208
- Serialization — 209
- Files — 211
- Summary — 212

Chapter 15: Embedding User Interface Components in Server-side Classes — 213
- A Typical MVC System — 214
- The Problem — 217
- The Solution — 218
- Which Approach Should We Use? — 221
- Summary — 221

Chapter 16: Tests Involving Databases — 223
- A Uniform Approach to Accessing the Database — 224
- Persistence Testing — 228
- Database Management — 229
- Summary — 231

Chapter 17: Function Tests — 233
- Specification of the Tests — 233
- Implementation of the 'DeleteCase' Test — 237
- Tests Involving Multiple JVMs — 240
- Multiple JVMs with GUI Components — 242
- Use of a Function Test as a Tutorial — 247
- Testing a Web Service — 251
- Summary — 256

Chapter 18: Load Testing — 257
- What to Test — 258
  - Overnight 'Housekeeping' Takes Too Long — 259
  - Deleting Cases Takes Too Long — 259
  - The BMD Server is Too Slow to Start — 260
- Measuring Time — 260
- Measuring RAM Usage — 264
- The Load Tests for LabWizard — 265
- Profilers and Other Tools — 266
- Summary — 267

Table of Contents

Chapter 19: GrandTestAuto — 269
What is GrandTestAuto? — 269
Unit Test Coverage — 272
Advantages of Using GTA — 273
Getting Started — 274
Testing Overloaded Methods — 277
Testing Protected Methods — 279
Extra Tests — 280
Classes That Do Not Need Tests — 280
Day-To-Day Use of GrandTestAuto — 281
    Running Just One Level of Test — 281
    Running the Unit Tests for a Single Package — 281
    Running the Unit Tests for a Single Class — 282
    Running the Tests for a Selection of Packages — 282
    Package Name Abbreviation — 283
    Running Tests for a Selection of Classes Within a Single Package — 283
    Running Individual Test Methods — 283
    Running GTA From Ant or CruiseControl — 284
    GTA Parameters — 284
Distributed Testing Using GTA — 285
    How it Works—In Brief — 285
    A Distributed Testing Example — 286
Summary — 287

Chapter 20: Flaky Tests — 289
A Flaky 'Ikon Do It' Unit Test — 289
Writing Reliable Tests — 294
Dealing with Flaky Tests — 296
Diagnostic Tools — 297
Tests That Do Not Terminate — 297
    Non-Daemon Threads — 298
    Remote Objects — 298
    Server Socket Still Waiting — 299
    Frame Not Properly Disposed — 299
Summary — 300

Index — 301

Preface

This book summarizes twenty years of experience testing software.

For the past decade, we have been the principal developers of **LabWizard**, which is the gold standard in Knowledge Acquisition tools and is used around the world to provide patient-specific interpretations of pathology cases. LabWizard is a very complex suite of software involving a server program, multiple remote client programs, interfaces with lab systems, and lots of internal processes.

In spite of this complexity, the software has been developed and maintained by a very small team with limited time and resources. We believe that our approach to testing, which we call **Extreme Testing**, has been central to our success. Extreme Testing has these key points:

- **Complete Unit Test Coverage:** All public classes must be thoroughly unit-tested.
- **Complete Requirements Test Coverage:** Each software requirement must be tested with an application-level test.
- **Test First:** When a bug is reported, a test that demonstrates the bug must be written before an attempt to fix the problem is undertaken.
- **Automation:** All of the tests must be run automatically.

This book is about why we have converged on this testing strategy, and how we actually implement it. In particular, we look at how to automatically test user interfaces, the help system, internationalization, log files, spreadsheets, email, web services, tests involving multiple JVMs, and a host of other things.

What This Book Offers

Above all, this book is a practical guide to testing Java software.

A number of books on Java testing have appeared in the last few years. In general, these books introduce the reader to JUnit, a commonly used test platform, and then run through some examples in which simple classes are tested. Additionally, they may introduce a few other tools for testing databases, dynamic web pages and so on. However, when it comes to testing user interfaces and other complex modules, these books draw a blank, and it is easy for the reader to get the impression that such testing is either not possible or is just too difficult to be worth the cost. We show how easy and worthwhile it is to automate these tests.

Another area covered in the present work, but absent from others, is that of application-level testing, which is the proving of our software against its requirements, item-by-item. These 'function tests' as we call them, are enormously useful and reassuring when we are in customer support mode.

For example, recently a customer rang us to report that blank lines in LabWizard reports are not appearing when the report is viewed in his lab's Laboratory Information System (LIS). Is this a LIS fault, or a LabWizard fault?

We actually have a function test class that does the following:

- Starts a LabWizard server in its own JVM.
- Starts a client program with which a Knowledge Base is built.
- Adds rules such that cases receive interpretations that have blank lines.
- Runs a LIS simulator that sends a case for interpretation.
- Checks that the interpretation received by the LIS simulator contains the appropriate formatting codes for the blank lines.

This test answers the question in LabWizard's favor. If we did not have this function test, we'd have to search through and understand a lot of code to respond to our customer.

By providing an approach to application-level tests, unit tests for difficult areas, and so on, we hope to prove that adopting an approach like ours will save time and money and make life as a developer less frustrating and more rewarding.

Every programmer interested in thoroughly testing their software should read this book.

Any programmer who is *not* convinced about the need to have automated tests as a part of their build process should also read this book. We hope that our experience with LabWizard, and the tools we provide, will change their opinion.

One way of getting more of an idea of what this book is about, is to consider the reasons why some developers don't write tests as one of their main day-to-day activities. Let's have a look at the things that come between programmers and the tests they should be working on.

Roadblocks to Testing

It's been over 25 years since G. Myers' classic *"The Art of Software Testing"*, so we know the principles. Every month we hear of some major software failure, so we know it's important. So how is it that any software is not tested effectively? Well, here are the roadblocks.

Roadblock 1: Last on the Waterfall

Whether we believe in longer or shorter development cycles, the "Waterfall" model still pervades our thinking and planning.

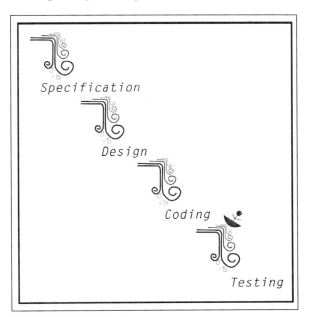

Where does testing appear in this process? Last, and at the bottom of the heap! And what things are done last? The least important things!

In this model, testing is divorced from specification, design, and coding. It is something we know we've got to do at some stage, but we'd rather not do just now—a necessary evil we put off for as long as possible.

With this mindset, when the development schedule starts looking ugly, and the development resources start looking thin on the ground, which activity is likely to get the chop?

Roadblock 2: Testing should be Done by Someone Else

We software engineers see our role as developers, which primarily means analysts, designers, and coders. Testing is something else entirely, and many programmers think that it should be done by a separate test team. Even Myers explicitly stated that we developers shouldn't test our own code as we can never see our own bugs.

Roadblock 3: It's Too Hard to Test User Interfaces

A third common belief is that testing is laborious when there is a user interface involved. In particular, any automated user interface test is seen as both very hard to set up and inherently unreliable due to the different appearance of components on different environments. How can we reliably position the mouse to click in a component, to activate a button, or to get focus to a tab? Finally, even if we do manage all this, when we are trying to test for some display, it's hard to get access to what is being drawn on the screen without putting a lot of code into our application that's only there for testing.

For these reasons, user interface testing is often done by manually following scripts, as this example from a well-known blog site suggests (see `http://software.ericsink.com/Iceberg.html`):

> ...Then I wrote some GUI test procedures for the new menu command. We keep documents which contain step-by-step instructions to verify that each feature in Vault works properly. (We also do a lot of automated testing below the GUI level, but that doesn't really apply in this example.)

Roadblock 4: It's Too Hard to Develop and Maintain Test Code

This myth is that thoroughly testing our software would take a tremendous effort, much more effort than writing it in the first place. We simply haven't got the time to do it. And even if the tests were written, the overhead of maintaining the tests once the software changes would again be way too high. Better to let the QA team do all the testing they can, and let the beta sites and early adopters find the rest of the bugs.

Roadblock 5: It's Too Expensive to Run Tests Often

To justify the investment in writing test code, the tests would have to be run frequently. But our software is only integrated every few months, just before the next release is due. This integration takes a lot of effort, and a team of people have to get involved. We simply couldn't afford to do this more frequently just so we could run our tests.

Roadblock 6: It's More Interesting to Develop "Real" Code

As highly creative developers, we find writing tests really boring. And by the way, that's not what we're paid to do. Testing is a lower task that should be done by others less qualified than ourselves, and less well paid.

In this book we will debunk the myths and work around (or just drive straight through!) the barriers to testing. In doing this, we will always be guided by practical considerations of what time-constrained programmers can realistically achieve.

Overview

Here is a very brief summary of each chapter.

Chapter 1: We explain the Extreme Testing approach—what needs to be tested, who is to write the tests, when they are to be written, what language they should be written in, and so on.

Chapter 2: We develop the unit tests for a non-GUI class. This shows the bootstrapping approach where tests, as far as possible, are written first. We also discuss Design by Contract.

Chapter 3: We discuss the management of test classes and test data.

Chapter 4: We present `Cyborg`, which is a higher-level wrapper for Java's `Robot` class. This is our most basic tool for user interface testing.

Chapter 5: We introduce a tool for managing and testing user interface text. This tool will make it much easier to write the 'wrapper' classes presented in Chapter 7.

Chapter 6: We discuss a design pattern that allows our user interface components to be individually tested.

Chapter 7: We introduce one of the main techniques for user interface testing, which is to build a 'UI wrapper class' for each component that needs testing. We illustrate this technique with a real example.

Chapter 8: We discuss some very important techniques for testing Swing applications. These techniques are: how to set up user interfaces in a thread-safe manner, how to find the objects representing particular components on the screen and how to interrogate a component for its state.

Chapter 9: We look in detail at the thorough testing of a simple component. This brings together a lot of the techniques we have developed in earlier chapters.

Chapter 10: We show code for testing commonly used Swing components including frames, lists, tables, menus, popup menus, combo boxes, progress bars, sliders, spinners, and trees.

Chapter 11: We introduce tools for creating and testing the JavaHelp system for an application.

Chapter 12: We examine some of the tools that make testing highly multi-threaded systems reasonably straightforward.

Chapter 13: Log files are important aspects of most server software, either for auditing or debugging. Here we look at how to test the logs produced by our software.

Chapter 14: Email, particular file formats such as spreadsheets, serialization, and so on are commonly used mechanisms for our software to communicate with other systems. Here, we look at ways of testing that these work.

Chapter 15: We show how to avoid the common problem of having too great a distinction between server and user interface classes.

Chapter 16: We introduce an approach to database management that facilitates testing.

Chapter 17: We look at how to specify and write the tests for our fully integrated system. Such tests typically have separate JVMs for the client and server components. We demonstrate our approach with a couple of the LabWizard function tests. We also look at how to test web services.

Chapter 18: We discuss how load tests fit into an automated test system, and how to write effective load tests.

Chapter 19: We discuss the advantages of using GrandTestAuto. We look at how it can be configured to run a small or large selection of tests, as we require in our day-to-day coding. We also show how we can distribute tests over a number of different machines.

Chapter 20: We look at what to do about tests that occasionally fail or do not terminate.

Examples and Code

This book is built around three example programs: **LabWizard**, **Ikon Do It**, and **GrandTestAuto**.

LabWizard

LabWizard is a set of tools that allow experts such as pathologists to capture their knowledge in an executable form—a "Knowledge Base". The underlying technology, called Rippledown Rules, was initially developed at the Garvan Institute of Medical Research and the University of New South Wales, and has been enhanced for commercial deployment at Pacific Knowledge Systems, where the authors work.

Each deployment of LabWizard has a server that persists and backs up the Knowledge Bases, manages user accounts, and communicates with the Laboratory Information System (LIS). There are several different flavors of LabWizard-LIS interface. At some sites, there is a shared directory that is polled for case files. At others, the LIS establishes a socket connection to LabWizard over which case data is sent. In our most recent deployments, the LabWizard server has a web service that the LIS queries.

There are several LabWizard client programs. The most interesting and novel is the Knowledge Builder:

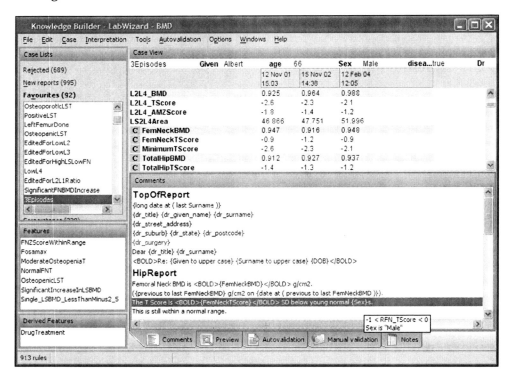

Pathologists use the Knowledge Builder to write the rules that encode their expertise. The beauty of LabWizard is that the expert does not know that they are building rules, and they do not need any programming experience. Their task is simply to look at cases and add, remove or replace comments as necessary, providing justifications for these changes in the form of natural language conditions. Behind the scenes, the software is working out the possible unwanted side-effects of the rule being added. The user prevents deleterious side-effects by adding stronger conditions to the rule they are working on.

Apart from the Knowledge Builder client, there is a Validator, which is the tool for reviewing interpretations. If a case is found to have been given an incorrect report, it is automatically queued to the Knowledge Builder where it will trigger corrective rule building.

There are several other client programs for specific tasks, such as translating a Knowledge Base into several languages, site administration, and other things. All of these clients communicate with the server using Java RMI. Of course, many different client programs can operate at once.

As this very brief description shows, LabWizard is extremely complex software involving novel algorithms and is a mission critical system in the labs in which it is deployed. Our Extreme Testing methodology has evolved to meet the needs of developing such software.

Throughout this book, screenshots of LabWizard and snippets of the code are used to illustrate various points.

Ikon Do It

Because LabWizard is proprietary software, and because it is extremely complex, it is not always ideal as a source of example code. Instead, we use Ikon Do It, which is a bitmap drawing program used for producing Portable Network Graphics (PNG) files that are suitable for use as program icons. This is a stand-alone program, much simpler than any of the LabWizard clients, but complex enough to provide good examples of automated unit tests of a user interface. The complete source and test code for Ikon Do It are provided. The following figure shows an icon being drawn with Ikon Do It.

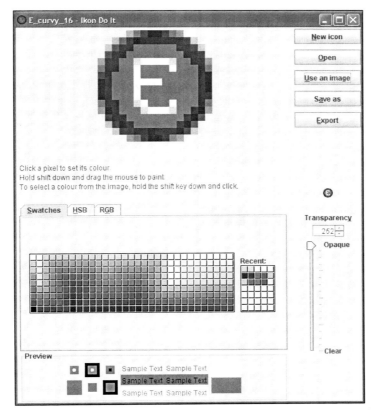

GrandTestAuto

Our preferred testing platform is GrandTestAuto (GTA). GTA is like JUnit on steroids—it automatically runs unit tests, application-level tests, and load tests, and checks that each class is properly tested. As well, GTA can be used to distribute tests across a number of machines on a network.

GTA itself is a good source of examples of automated tests, and these are used throughout the book and are available with the source code. The requirements and test specifications are provided too, as examples of the way the requirements drive the application-level testing, as discussed in Chapter 1 and Chapter 17.

Source code

The text of this book is only a part of the story. The other part is some 200-odd java files of source code that readers are welcome to use in their own projects.

What You Need For This Book

We assume that the readers of this book have some experience with Java programming and basic design patterns and are familiar with the Ant build tool. Five of the chapters are about testing Swing user interfaces. To get the most from these, some experience of Swing programming is required. Other chapters deal in detail with particular aspects of Java programming, for example, Chapter 5 is mostly about internationalization, and Chapter 11 concerns the JavaHelp API. Readers without knowledge of these topics can use the Java Tutorial or some other resource to get up to speed, or simply move on to the next chapter. By browsing (or even better, running) the source code, readers should be able to follow any of the code samples in the book easily.

Who Is This Book For

This book is for Java developers. It will primarily be of interest to those working on client-server or desktop applications, but all of the theory and some of the practical examples will help web-based developers too.

Conventions

In this book, you will find a number of styles of text that distinguish between different kinds of information. Here are some examples of these styles, and an explanation of their meaning.

Code words in text are shown as follows: "We can include other contexts through the use of the include directive."

A block of code will be set as follows:

```
robot.clickButton( buttonInFrame );
WaitFor.assertWaitToHaveFocus( buttonInDialog );
robot.escape();
noErrorFound = frameCursorType() == Cursor.DEFAULT_CURSOR;
```

When we wish to draw your attention to a particular part of a code block, the relevant lines or items will be made bold:

```
public boolean runTest() throws IOException {
   creationTimer = new Stopwatch();
   runTimer = new Stopwatch();
   //a51:  27 classes in all.
   runForConfiguredPackages( Grandtestauto.test51_zip );
```

Any command-line input and output is written as follows:

`org.grandtestauto.GrandTestAuto GTASettings.txt`

New terms and **important words** are introduced in a bold-type font. Words that you see on the screen, in menus or dialog boxes for example, appear in our text like this: "clicking the **Next** button moves you to the next screen".

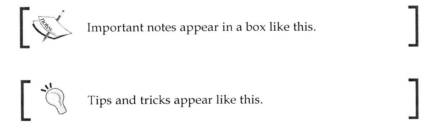

[Important notes appear in a box like this.   ]

[Tips and tricks appear like this.   ]

Reader Feedback

Feedback from our readers is always welcome. Let us know what you think about this book, what you liked or may have disliked. Reader feedback is important for us to develop titles that you really get the most out of.

To send us general feedback, simply drop an email to feedback@packtpub.com, making sure to mention the book title in the subject of your message.

If there is a book that you need and would like to see us publish, please send us a note in the **SUGGEST A TITLE** form on www.packtpub.com or email suggest@packtpub.com.

If there is a topic that you have expertise in and you are interested in either writing or contributing to a book, see our author guide on www.packtpub.com/authors.

Customer Support

Now that you are the proud owner of a Packt book, we have a number of things to help you to get the most from your purchase.

Downloading the Example Code for the Book

Visit http://www.packtpub.com/files/code/4824_Code.zip to directly download the example code.

The downloadable files contain instructions on how to use them.

Errata

Although we have taken every care to ensure the accuracy of our contents, mistakes do happen. If you find a mistake in one of our books—maybe a mistake in text or code—we would be grateful if you would report this to us. By doing this you can save other readers from frustration, and help to improve subsequent versions of this book. If you find any errata, report them by visiting http://www.packtpub.com/support, selecting your book, clicking on the **let us know** link, and entering the details of your errata. Once your errata are verified, your submission will be accepted and the errata are added to the list of existing errata. The existing errata can be viewed by selecting your title from http://www.packtpub.com/support.

Questions

You can contact us at questions@packtpub.com if you are having a problem with some aspect of the book, and we will do our best to address it.

1
What Needs Testing?

The aim of testing is to find errors in our software.

Any deviation from the required behavior of an application is an error. The 'required behavior' is defined by the so-called **user stories** in Extreme Programming (XP), or by a requirements specification in more traditional methodologies. Additionally, there will be implicit requirements on usability, reliability, and scalability. These may be derived from company or industry standards, customer expectations, user documentation, and various other sources.

Clearly, we need tests to prove that our software satisfies these formal specifications. However, it would be completely unrealistic to think that we can thoroughly test our application by testing it against each of these high-level requirements or company standards, and not testing the components comprising the application. This would be like attempting to production-test a new car simply by driving it without having first tested each component in isolation, such as the brake system. Flaws in a component may manifest themselves in scenarios that we could never have imagined beforehand, and so would never have designed an application-level test case for.

Therefore, in terms of testing infrastructure, we need to have at least two views of our application:

- The **Unit** or **Module** view, where "unit" or "module" refers to the smallest compilation component. In Java, this is a class.
- The **Application** view, where "application" refers to the complete set of classes providing the application. The application will be made up of our developed components, the standard Java class libraries, plus libraries of third-party components, executing on one or more JVMs.

Let's now look at some of the broader aspects of testing at the unit and application levels. The nuts and bolts of how to implement and organize our tests at these levels will be dealt with the later chapters. First of all we will take a look at an example.

An Example

A good example of the need for proper unit testing came up recently with LabWizard.

For a LabWizard clinical Knowledge Base, the Pathologist uses a tool called the Validator to review reports generated by the Knowledge Base, make changes if necessary, and then release them to the referring doctor:

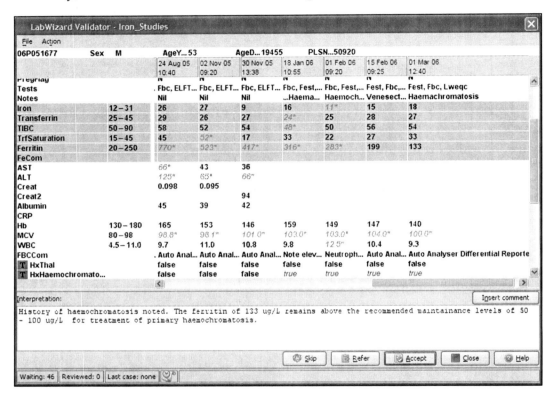

The main screen here is an instance of a class called **CaseViewer**. The patient demographics and test history are shown in the table. The interpretive report is shown in the text area below the table.

After reviewing the interpretation, and making any necessary changes to it, the clinical reviewer activates the **Accept** button, and the report is sent back to the Laboratory Information System where it is automatically released to the referring doctor.

Note that the **Accept** button has the focus by default. This is in fact a serious usability bug. There were several instances of a pathologist unwittingly pressing the space bar or the *Enter* key, and thereby releasing an inappropriate report. Rather, we would want the activation of the **Accept** button to be a very deliberate action, that is, by using either the mouse or the mnemonic.

The fix for this bug was to ensure that the focus is on the **Skip** button whenever a new report is presented for review. The test for this new behavior took only a matter of minutes to write, and runs in just a few seconds. Here it is:

```java
/**
 * STR 1546. Check that the accept button does NOT have
 * the focus to begin with, nor after a case is approved.
 */
public boolean acceptButtonDoesNotHaveFocusTest() {
    init();
    //Queue some cases and wait till the first one is showing
    cvh.setNumberOfCasesToBeReviewed( 5 );
    waitForCaseToBeShowing();
    //Ensure that no cases have been approved yet
    waitForNumberOfApprovedCases( 0 );
    //Pressing the space bar should skip the case, not approve it
    robot.space();
    //Give plenty of time for a case to be sent to the LIS
    //if it were going to be sent.
    TestSetup.pause( 1 );
    //Check that still no cases have been approved.
    waitForNumberOfApprovedCases( 0 );
    //Now approve a case using the mnemonic,
    //and check that the focus shifts off the Accept button, back
    //to the Skip button
    mn( SimpleMessages.ACCEPT );
    waitForNumberOfApprovedCases( 1 );
    //Pressing the space bar should again skip the case
    robot.space();
    //Check that no more cases have been approved.
    TestSetup.pause( 1 );
    waitForNumberOfApprovedCases( 1 );
    cleanup();
    return true;
}
```

The structure and specifics of this test will only be clear by the end of Chapter 10. But even at this point, it should be apparent how simple and easy it can be to write user interface tests. Note also the convention for referencing the Software Trouble Report (STR) describing the bug in the comment for this test method.

Another point from this example is the necessity for such tests. Some developers want to believe that testing user interfaces is not necessary:

> *Of course GUI applications can (and should) be unit tested, but it isn't the GUI code itself that you want to test. It's the business and application logic behind the GUI. It's too easy to just say "don't unit test GUIs" because it's difficult. Instead, a simple separation of logic from GUI code means that the stuff which should be tested becomes easy to test; and the mechanical stuff (Does the UI freeze when I click this button? Is the layout correct?) can be left for visual inspection. (You do test your app by running it from time to time, don't you? And not just by waiting for a green bar.) If the UI itself doesn't behave, you'll quickly know about it. But if the application logic is wrong, you might not find out until the application has been released, unless you've got a decent set of unit tests. (From* **Don't unit test GUIs** *by Matt Stephens, the Agile Iconoclast. See* `http://www.regdeveloper.co.uk/2007/10/22/gui_unit_testing/`*).*

With this particular bug, it really was the GUI that we wanted to test. We could do a manual test to ensure the default focus was not on the **Accept** button, but it would definitely take longer than the automated one that we've got. Why have people doing anything that a computer can automate? Furthermore, focus behavior can change unexpectedly as components are added or removed from a container. So a manual test in a previous release of our software would give us little confidence for subsequent releases.

The unit test is an appropriate place for tests, such as this, of the specific behavior of a component. Although we could test these things at the application level, such tests would be harder to write, more difficult to understand and maintain, and slower to run than the equivalent unit tests.

What Classes Do We Test?

Our experience is that in software Murphy's Law holds supreme: *any class without a test is guaranteed to contain bugs.*

So to the question, "What classes should be tested?", there is only one satisfactory answer:

 Extreme Testing Guideline: Every public class requires an explicit unit test.

By "explicit unit test", we mean a test that creates an instance of the class and calls methods on that class directly. Whilst most software engineers will assent to this guideline in theory, it is surprising how difficult it can be to implement this as a work practice. Let's briefly consider why.

The most common objection is that there's not enough time to write the unit tests, either for existing or new classes. That is, there is a commercial imperative to write new code as quickly as possible, rather than "waste" time writing tests. In our experience of more than 20 years with many different types of software projects, this has never been a valid argument.

The cost of fixing a bug rises exponentially with the length of time it takes to find it. In particular, bugs found once the application has been released can typically be ten times more expensive to fix than bugs found while developing the class. For example, the LabWizard, "Accept button focus bug" took a matter of minutes to test for, and fix. But it took more than an hour of our time to resolve the problems it caused, not to mention the concern of the Pathologist, who had inadvertently approved an inappropriate report! ("*The Economic Impacts of Inadequate Infrastructure for Software Testing*", National Institute of Standards and Technology, May 2002, `http://www.nist.gov/director/prog-ofc/report02-3.pdf` is a worthwhile read). Bugs in deployed software effectively plunder development resources for use as support resources.

Furthermore, bugs that are found late in the development life cycle, or even worse, after releasing the software, have a very negative effect on team morale. We all have had the unfortunate experience of releasing a version of an application, then eagerly anticipating the development of exciting new features in the next release, only to be stopped dead in our tracks by customer-reported bugs that require us to revisit the version we just released and had thought we were finished with. The exact opposite is true for bugs found during unit tests—each one we find gives us a sense of achievement.

Another common argument is that certain classes are too trivial to require a test. Our response is that, the simpler a class, the simpler will be its test, so what's the problem? That argument also misses the point that unit tests are not just testing the current version of the code, they also test future versions. Simple classes are likely to become more complex as a project evolves, and changes to a class can all too easily introduce errors. This is particularly true if the original authors of a class do not make all the changes themselves. Putting unit tests in from the very beginning makes this evolution a safer process and pays for the cost of testing up-front. For an excellent rebuttal to many of the arguments against testing early and often, see "*Myths and realities of iterative testing*" by Laura Rose, IBM developerWorks.

The converse may also be argued, that certain classes are too complex to be tested. However, complex classes are the ones that are least likely to be understood, most likely to contain errors, and most difficult to maintain.

Unit testing address these issues by:

- providing a specification of the class' behavior, hence enhancing our understanding
- increasing our chance of detecting errors, and
- enhancing maintainability by providing a behavioral benchmark against which to measure the effect of further changes to the class, or changes to the way it is used.

It may be that the class in question is indeed overly complex. For example, it does not represent a single well-defined abstraction. In this situation it may well need to be redesigned to reduce its complexity. One of the benefits of unit testing above the mere fact of detecting errors, is the increased discipline that it brings to design and implementation.

Of course, in Object Oriented software, a single class can contain a lot of sub-objects that are themselves quite complex. For example, a LabWizard `CaseViewer` contains classes to interface it to the LabWizard server, and a Swing `Timer` polling for more cases, as well as all the GUI components and associated actions. Despite this complexity, we have a unit test class for `CaseViewer` that tests every GUI component and every exposed method, just as we do for our most basic utility classes. This unit test has been invaluable in preventing bugs from creeping into one of our most important user interface components. Based on our experience, we'd say that a class that is seen as too complex to test, is simply too complex.

We will talk about testing legacy code later in this chapter.

Some programmers will argue that a class does not have much meaning outside its package, and so can't be tested independently. Our response is that if the class is tightly coupled within a package, it should be made private or package visible, not public, in which case it need not be explicitly tested, as we will discuss below.

Finally, some programmers believe that their code is bug-free, so why should they bother with tests? Of course, this claim is wrong in most cases. Even if some class is indeed bug-free at some point in time, there is no guarantee that it will remain bug-free in the next release after changes have been made to it, or to the classes it depends on, or to the classes that depend on it.

In fact, there is an even more compelling reason why all classes must be unit tested. If we adopt the XP test-first approach to unit development, which we strongly endorse, the test class must be developed first, that is, before the production class.

Test First—Always!

> **Extreme Testing Guideline:** We should implement our unit test classes before we implement our production classes, to the extent that this is possible.

This guideline should be read as being an unattainable goal. Some compromise and common sense is needed in its interpretation. We cannot really write a unit test for a class before the class itself, as the unit test will not compile. The best we can do in practice is to write a small amount of the production class, then test what we have written. We might have stub implementations of the production class methods, so that our code compiles. In this sense, our tests are ahead of our production code. Our aim is to always keep the test code as far ahead of the production code as is practical. An example of the test-first approach is given in the next chapter.

There are several reasons for this guideline.

Firstly, the test class provides an unambiguous specification of the corresponding production class. Even after following the necessary software engineering practices of requirements and design analysis, we are always surprised when we start writing our test class at just how many specification issues are in fact still undecided. All these issues need to be resolved by the time our test class is complete, and so the test class becomes the definitive and concrete expression of the production class' specification.

Secondly, writing our test class first means that we know when to stop implementing our production class. We start with an absolutely trivial implementation, and stop once all our tests pass. This discipline has the benefit of keeping 'code bloat' down. As developers, we find it too easy to add 'convenience' methods to a class that might be used later on in the lifecycle of a project. It's tempting to think that these methods are so simple that even if they're never used, there's no harm done. But such methods actually do cost something. They cost a few minutes to write initially, and they make a class harder to understand. If they are not maintained properly, the behavior of such a method might change over time and, years later, when eventually used, they might cost time by not actually working as expected. Forcing developers to test every method they write minimizes the number of unused methods, simply because nobody really wants to write tests that they don't have to.

If all our tests pass, but there is some aspect of our class that we think is not yet quite right, then our tests are inadequate. Therefore, before going any further with our production class, we first enhance the test class to check for the desired behavior. Once our production class fails the new test, we can return to the production class, and implement the desired behavior—and our implementation is complete when once again the tests pass.

Finally, if we attempt to write a test class after the production class has been developed, the chances are that there will be aspects of the production class that will need to be changed to make it more easily testable. It may even need to be substantially redesigned. Writing the test class before the production class is therefore a much more efficient use of our time, as this additional effort in rework or redesign is largely eliminated.

The test-first approach also applies in the maintenance phase of a class. Once a bug is scheduled to be fixed, the test class must first be modified to reproduce the bug, and so fail. The production class can then be modified until the test passes again, confirming that the bug has indeed been fixed.

This is precisely what we did with the test method `acceptButtonDoesNotHaveFocusTest()` shown earlier. When first implemented, this test failed half-way through at the check:

```
...
robot.space();
TestSetup.pause( 1 );
waitForNumberOfApprovedCases( 0 );
...
```

showing that the bug had been reproduced. If we had modified the test code after the bug had been fixed in the production `CaseViewer` class, we could never be sure that the test would have reproduced the error, and hence we could never be sure whether the fix we applied to the production class was adequate.

[**Extreme Testing Guideline**: We should modify our tests to reproduce a bug before we attempt to fix it.]

What Classes Don't We Test?

In general, we don't recommend explicit testing of private or package visible classes. There are several reasons for this.

Firstly, any private or package-visible classes must eventually be used by public classes. The unit tests for the public classes can and must be made thorough enough to test all the hidden classes. There is no compelling need to explicitly test hidden classes.

Secondly, the separation of production code from test code is highly desirable from many perspectives, such as readability of the production code, ease of configuration control, and delivery of the production code. Test code for a class can often be an order of magnitude larger than the production code itself, and often contains large sets of configured test data. We therefore like to keep our test classes in explicit test packages so that they do not clutter up the production packages. This clutter can, of course, be removed from our delivered software, but it does make the production classes more difficult to navigate and understand. The separation of the production package from its test package means that the private and package-visible classes cannot be directly accessed by the test packages (though there are workarounds, which we'll look at).

Thirdly, private and package visible classes will normally be tightly coupled to the owning or calling classes in the package. This means that they will be difficult, if not impossible, to test in isolation.

Finally, within a public class, the decision to create private classes or private methods can at times be fairly arbitrary, to aid readability for example. We don't want to restrict the ability to do this sort of refactoring with a requirement that all such restructuring necessitates more testing.

A simple example of this is in our `CaseViewer` class, which has a Swing `Timer` for polling the server for cases. We decided to make the action listener for this `Timer` a private class, for the sake of readability:

```
private class TimerActionListener implements ActionListener {
        public void actionPerformed( ActionEvent ae ) {
            //reload the case and interpretation
            rebuild();
        }
    }
```

Other classes that may not need an explicit test are renderers. For example, the table cell renderer used in the `CaseViewer`.

Private and package visible classes like this do, however, need to be implicitly tested, and this will be done by white-box testing the public classes. That is, we design our tests taking into account the internal structure of the public classes, not just their specification.

We can use code coverage tools to check whether we have really exercised all our hidden code by the explicit unit tests of our public classes. We will look at an example of this later.

If a private or package visible class is too complex to be implicitly tested easily, it might be pragmatic to make it public so that it can be adequately tested. Clearly, this should only be done as a last resort.

[**Extreme Testing Guideline**: It's OK to make a class public if it can't be tested in any other way.]

A lot of developers are horrified at the thought of increasing the exposure of a class for any reason. The principal objection is that by exposing a class in our public API, we may be forever committed to maintaining it in its visible state. Whether or not this is a problem in practice will depend on the project being developed. If we are publishing a class library, the objection might be valid. But for most other projects, there really is no reason not to do this. In the situations where it would be really wrong to expose a class, we can write tests that use reflection to call methods and constructors of hidden classes, but this approach has its own problems.

As well as hidden classes, we do not consider it necessary to explicitly test auto-generated classes. Simple `Enums` fall into this category, having methods `values()` and `valueOf()` that are generated by the Java compiler. `Enums` that define other methods should be tested.

What Methods Need Testing?

Granted that we need to test each public class, what needs to be tested in a class? The specifications of the behavior of a class are the public and protected method specifications, plus any controls used for event-driven input or output.

[**Extreme Testing Guideline**: The unit test for a class must test all public and protected methods and constructors and all user interface components provided by the class.]

The reasons are precisely the same for requiring that all public classes be tested.

Included in this guideline are the accessible class constructors. We need to test whether these create an object that is in a consistent and expected state.

A lot of developers are of the opinion that there is no need to explicitly test public "getter" and "setter" methods of a class:

> *JUnit convention is that there is a test class created for every application class that does not contain GUI code. Usually all paths through each method in a class are tested in a unit test. However, you do not need to test trivial getter and setter methods. (See* `http://open.ncsu.edu/se/tutorials/junit/`.*)*

Our view is that the existence of paired getters and setters, if all they do is set or expose a variable, is usually an indication of poor design. They expose the internal state of the object as effectively as a publicly visible variable. (For more on this, see JavaWorld article "*Why getter and setter methods are evil*" by Allen Holub, September 2003.)

So the argument that "getters" and "setters" do not need tests is 'not even wrong', as they say. If there are good reasons for having such a method, then a test is warranted. The implementation could evolve from simply returning a value to returning a more complex calculation, for example. Without a test, bugs could easily be introduced.

As mentioned earlier, a great and perhaps unexpected benefit of the extreme approach to testing is that only the methods that are absolutely necessary ever get written. By forcing ourselves to write unit tests for all methods, we keep ourselves focused on the task at hand—producing lean, but high-quality code. At the end of this chapter, we'll see just how effective this has been at keeping the LabWizard code under control over the last ten years.

So, as a bare minimum, we will need to have a unit test method for each accessible method and constructor in a class. For some classes, this minimal amount of testing is enough, because the class' behavior is immediately implied by the public methods. Other classes have only a few public methods but a lot of code that is called from within a framework such as Swing, or from an internal thread. For example, a typical user interface class, such as the `CaseViewer`, might have a very long constructor that does a lot of layout and sets up some event handlers, plus one or two simple public methods. For such a class, each unit test method will correspond to either a behavior that needs testing or to an accessible method or constructor.

What Needs Testing?

For example, the `CaseViewer` class has 20 public methods, but its unit test class has over 60 test methods, a selection of which is shown here:

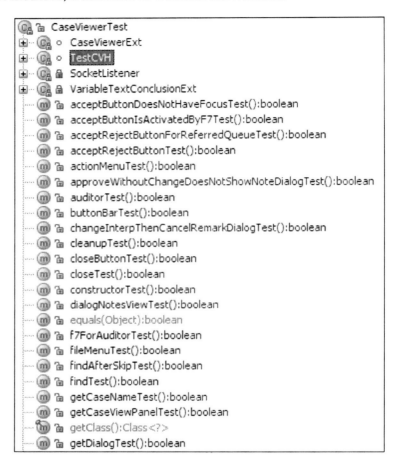

We will look at this in more detail in Chapter 7. Another example of functionality that needs to be tested, but does not relate directly to a public method, is the serializability of classes declared to implement the marker interface, `java.io.Serializable`. This issue is discussed in Chapter 14.

What Methods Don't We Test?

For precisely the same reasons as given for the private classes, we do not recommend explicit testing of the private or package visible methods of a class.

The following is an example of a very simple private method in `CaseViewer` that determines whether the user has changed the interpretive report of the case currently showing.

```
private boolean interpretationHasBeenChangedInThisSession() {
    return kase.differsFromInterp( interpretationText() );
}
```

Even though a one-liner, this was usefully implemented as a method to aid readability and to avoid copying that line in several places throughout `CaseViewer`.

Although not explicitly tested, these methods must be implicitly tested. For classes, there may be times when we have to expose a method (make it public or protected) just so that it can be adequately tested. We may also have to use reflection to change the accessibility of the method on-the-fly, as we describe in the next section. Again, this should be done only as a last resort.

Another situation that may arise is where there are public methods that are "paired", in the sense of performing inverse operations. Some example methods may be:

```
/** Serializes the object as an XML string */
public String toXML(){...}
/** Construct an object from an XML string */
public Object fromXML(String str){...}
```

It would make sense for some of the test cases for `toXML()` to use `fromXML()`, and in fact could equally well serve as test cases for `fromXML()`. Rather than just repeat these test cases in both methods, it may be sufficient to implement them just once, say in the tests for `toXML()`. That is, the method `fromXML()` would still need to be tested, but some of its test cases may be found in another test method.

Invoking Hidden Methods and Constructors

Java's access modifiers can be overridden using reflection. The key to this is the `setAccessible()` method of `java.lang.reflect.AccessibleObject`.

Consider a package with a package-private class called `HiddenClass` that has a package-private constructor and a package-private method that we want to test:

```
package jet.reflection_example;
    class HiddenClass {
        private long createdTime;
        HiddenClass() {
```

```
            createdTime = System.currentTimeMillis();
        }
        long createdTime() {
            return createdTime;
        }
    }
```

It is possible to explicitly test this class and have the test in a package separate from the class itself, if we are prepared to do a few tricks with reflection. Here is the test:

```
package jet.reflection_example.test;

import java.lang.reflect.*;

public class HiddenClassTest {
    public boolean runTest() throws Exception {
        Class klass = Class.forName(
                        "jet.reflection_example.HiddenClass" );
        Object instance = createInstance( klass );
        Object value = invoke( klass, instance, "createdTime" );
        System.out.println( "m.invoke() = " + value );
        return true;
    }
}
```

In the first line of `runTest()`, we find the `Class` object for `HiddenClass` using the `Class.forName()` method. Having found the class, we can create an instance of it using our `createInstance()` method, which seeks a no-arguments constructor, changes the accessibility of it, and invokes it:

```
Object createInstance( Class klass ) throws Exception {
    Constructor noArgsConstructor = null;
    Constructor[] constructors = klass.getDeclaredConstructors();
    for (Constructor c : constructors) {
        if (c.getParameterTypes().length == 0) {
            noArgsConstructor = c;
            break;
        }
    }
    assert noArgsConstructor != null :
            "Could not find no-args constructor in: " + klass;
    noArgsConstructor.setAccessible( true );
    return noArgsConstructor.newInstance();
}
```

The highlighted line makes the constructor usable, even if it is private, or package-private. To invoke the method we want to test, we search for it by name, change its visibility, and invoke it:

```
Object invoke( Class klass, Object instance,
            String methodName ) throws Exception {
    Method method = null;
    Method[] declaredMethods = klass.getDeclaredMethods();
    for (Method m : declaredMethods) {
        if (m.getName().equals( methodName )) {
            method = m;
        }
    }
    assert method != null :
        "Could not find method with name '" + methodName + "'";
    method.setAccessible( true );
    return method.invoke( instance );
}
```

This example shows how we can circumvent the Java language access controllers, and run hidden methods in our tests. The pitfalls of working this way are that the code is harder to understand and harder to maintain. Changes to method names or signatures will break the tests at runtime, rather than at compile time. Whether or not we take this approach depends on the project we are working on—is it an absolute requirement that as few classes and methods are exposed as possible, or can we take a more pragmatic approach and expose classes and methods for explicit testing in those rare situations where implicit testing is not adequate?

Unit Test Coverage

Our preferred testing harness, GrandTestAuto, enforces the testing policy we have developed so far in this chapter. This policy requires that all accessible methods and constructors of all accessible classes be explicitly unit-tested. By using this tool, we know that the exposed programming interface to our code is tested.

Another aspect of test coverage is checking that our code is thoroughly exercised by our tests. This can be measured using tools such as Cobertura (see `http://cobertura.sourceforge.net`) and EMMA (see `http://emma.sourceforge.net`). Here, we will give an example using EMMA.

The EMMA tool works by installing a custom class loader that instruments the classes we're interested in. The tests are run, and as the JVM exits, a series of HTML reports are written that indicate the level of coverage achieved. For example, running the unit tests for the CaseViewer class only confirmed that the tests had exercised all 94 methods of the class (that is, including private methods) and 98% of the 485 source lines. The following figure shows the Emma coverage report for the CaseViewer unit tests:

COVERAGE BREAKDOWN BY SOURCE FILE

name	method, %	line, %
CaseViewer.java	100% (94/94)	98% (476.8/485)

A closer look at the report showed which methods had not been covered:

applyCustomisationsForValidator (): void	100% (1/1)	100% (28/28)
approveCurrentCaseAndClearTheCaseView (String, boolean): void	100% (1/1)	100% (3/3)
buildNotesArea (): JTextArea	100% (1/1)	100% (9/9)
buttonBar (): JComponent	100% (1/1)	100% (1/1)
caseRefreshed (): void	100% (1/1)	97% (28/29)
cleanup (): void	100% (1/1)	78% (7/9)
clearCase (String): void	100% (1/1)	100% (16/16)
close (): void	100% (1/1)	100% (7/7)
closeButton (): JButton	100% (1/1)	100% (1/1)
createCaseViewAndInterpArea (): void	100% (1/1)	100% (34/34)
createMenusAndButtons (): void	100% (1/1)	100% (21/21)

Drilling down into the EMMA report showed us which lines of production code were not being exercised by the tests. The unexercised lines in both the `caseRefreshed()` and `cleanup()` methods were, in fact, gaps in our testing.

Who Should Implement the Unit Tests?

It is clear that if we adopt the test-first methodology, the developers of a new class must be the ones initially responsible for the development of the corresponding unit tests. There's simply no choice in this matter. The test class is too intimately connected with the specification and design of the production class to be done by anyone else.

Similarly, once the class and its unit test have been developed, and if changes need to be made for a subsequent release, the test class will be modified by the same developers who modify the production class.

What About Legacy Code?

We've had the experience of having to write unit tests for untested legacy code, long after the original developers have moved on, and we know what a frustrating task this is. An approach we would recommend is to develop the tests in an incremental fashion. That is, develop corresponding unit tests whenever a part of the legacy code needs to be modified. For example, if a method `x()` of an untested legacy class X needs to be modified, create a test class and methods to thoroughly test `x()`. But just use "`return true`" test stubs for the other methods of X that you don't need to modify. We shall see a working example of this later.

In this way, each iteration (in XP parlance, an iteration is a software release cycle) will result in more of the legacy code being adequately tested.

Where Does Integration Testing Fit In?

Using the XP methodology, each iteration provides a fully integrated (but not necessarily fully functional) version of the application. Furthermore, using an Object-Oriented design methodology, the higher level classes will have the role of integrating the lower level classes, normally by delegation or by containment relationships. Hence, the combination of both methodologies ensures that unit testing of the higher level classes will provide our integration testing.

We saw this earlier with the `CaseViewer` class in LabWizard. In the unit test for this class, we are testing objects that are almost complete client applications. By virtue of the fact that these tests run at all, we are confident of the integrity of our class structure. The Validator is a particularly simple application, but unit tests verify the integration of our very complex applications and even of our server application.

Documentation of Unit Tests

Most unit tests can be adequately documented by the class and method comments in the unit test class itself. The test class naming convention (see Chapter 19) will ensure that there will be no ambiguity about which production class a test class refers to.

However, it may be that a class will require more than one test class. This most recently happened in LabWizard testing when significant new functionality was added to a class called `RuleTreeImpl`, which had not changed for several years. To effectively test the new method, for removing all rules associated with a certain kind of conclusion, we needed to set up rule trees in a way that was not easy to do within the existing unit tests. So we added a new test class called `ReportedConclusionRemovalTest` that did just this. In general, a descriptive name for the additional test class, plus comments at the point it is called, should suffice to identify its role.

Testing at the Application Level

The testing approach that we recommend at the application level is, perhaps surprisingly, very similar to the approach for unit-level testing presented so far. However, application-level testing has its own challenges in terms of specification and implementation, and we will deal with some of these issues in the later chapters.

Myers (G. Myers, *The Art of Software Testing*, Wiley, 1979, Chapter 6.) gives an excellent overview of the categories of higher level testing. They can be summarized as follows:

- **Functional Testing**, against the requirements specification.
- **System Testing**, against the user documentation, or company standards if not explicitly specified in the requirements specification. The sub-categories of System Testing include:
    - Performance, Stress, and Volume
    - Interface
    - Usability
    - Reliability and Maintainability
    - Security
    - Installation
    - Free play
- **Acceptance Testing**, against a contractual specification.

What requirements should be tested? As for unit testing, the only satisfactory answer is:

[**Extreme Testing Guideline**: All application-level requirements need to be explicitly tested.]

Each requirement should explicitly reference one or more test cases, normally defined in a test specification. Each test case should be a script that defines the test data to be used, steps providing input to the application (for example, via the user interface or from other software), and the expected outputs as unambiguous pass/fail criteria.

Most importantly, each test case should be implemented as a test class that can be executed without any manual intervention. To do this, we will need high-level "test helper" classes, which provide input to the application via keyboard, mouse, and data interfaces, and conversely, can read outputs from the application such as screen displays. These helper classes will be described in detail in the later chapters. In Chapter 17 we look in detail at some application-level tests for LabWizard.

The LabWizard Validator application now has 60 application-level tests, and that is one of the smaller applications within LabWizard.

In the specification and testing of LabWizard, almost all the automated application-level tests come from the functional specification of the software. For this reason, at Pacific Knowledge Systems, we use the term "function test" for "automated application-level test". We will use the term "function test" in the rest of this book.

Who Should Implement the Function Tests?

The test-first methodology again implies that developers must implement function tests. At this level, requirements are even more ambiguous than at the unit level, mainly because of what a requirement may leave unsaid. A comprehensive set of test cases is needed to rigorously define the requirements of a high-level feature before it is developed.

In an ideal scenario, an independent test team, or the customer, will provide a valuable source of additional test cases that the development team simply would never think of. After all, the test team's primary objective is to break the application. As developers, we often find it difficult to make sincere attempts to find fault with what we have produced, though this is a skill that we must acquire to be fully effective, test-driven, programmers.

Automated Test Execution

In the "bad old days", a unit test, if it existed at all, would be run at the time the unit was developed and included in the build. The test execution would be initiated by the developer, possibly by calling a method in the class that would call some test methods, or draw the component in a test frame that allowed for visual inspection and manual input. The test may never be run again, especially if that component did not change.

Function tests would be run just prior to the planned release of the next version — the last stage of the Waterfall. They would be executed manually against some test specification, normally by junior engineers in the team or dedicated test engineers. The results would be manually recorded.

This type of test execution is the antithesis of the Extreme Testing approach and is likely to result in a low-quality software release. With each new piece of functionality introduced, new bugs will be created, or old bugs will resurface.

How then should tests be executed?

[**Extreme Testing Guideline**: Unit and function tests must be developed within a framework that provides a fully automated execution environment.]

Examples of test execution environments are JUnit (http://junit.sourceforge.net) and GrandTestAuto. The latter is our preferred tool, and is the subject of Chapter 19.

There are several compelling reasons for insisting on fully automated tests:

- **Early error detection**: The running of the tests simply becomes another stage in the daily automated build procedure, guaranteeing that any error is picked up as soon as possible.
- **Regression**: It often happens that an error introduced into a modified class A only manifests itself in an interaction with some other class B. Running all the tests each day, rather than just running the test for the modified class A, for example, maximizes the chance of picking up these more subtle types of errors. That is, automation allows an effective regression test strategy.
- **Consistency**: Test results will be far more consistent than if executed manually, no matter how diligent and careful a tester is.
- **Coverage and depth**: With an automated methodology, the coverage and depth of the testing that is possible is far greater than anything that can be achieved manually. For example, a huge variety of test cases, often with only subtle differences between them, can be readily implemented so that some feature can be fully tested. This variety would be impractical if the tests were to be run manually.
- **Convenient reporting**: With an automated test methodology, test results can be conveniently reported in a summary fashion at any level of detail. In particular, the test report will quickly identify the set of failed tests.
- **A prerequisite for incremental development**: The effort involved in running all tests is a limiting factor on the amount of time one can devote to a development iteration. For example, if it takes two weeks to run all the tests, it's impractical to have development iterations on a three or four week schedule—the testing overhead would be far too great. Conversely, if the execution of tests can be done as part of the daily build cycle, development iterations can be very short.

There are of course, two types of application-level testing which can't be fully automated, almost by definition.

With free-play testing we need to execute unplanned test cases, so clearly, these test cases can't be automated, at least not the first time they are run. If some free-play testing does find an error then it makes sense to include this scenario as an automated test case for subsequent execution.

Similarly, whilst a large part of usability testing may be automated, for example, checking the tab order of components in a dialog, or checking that each item in a menu has a mnemonic, there will be elements of usability testing that can be done only by a human tester.

A Hierarchy of Tests

We only release software once all the tests for it have passed. In this sense, it does not matter in what order the tests are run. In reality though, we want to find errors as quickly as possible, both in our continuous build process, and in our final release testing.

As an example, here is how we organize the tests for the LabWizard product suite. This is not to be interpreted as the best way of doing things for all projects. It is simply an example of what has worked for us.

We organize the testing into three stages:

First are the **unit tests**. These cover every single class, including those complex ones that draw together lots of other classes to create significant functional units, even stand-alone products. As mentioned earlier, the very nature of object-oriented systems means that complete unit-testing achieves a great deal in the way of integration testing. The times taken by the unit tests vary from a few seconds to tens of minutes, depending on the complexity of the classes that they test. Unit tests for user interface classes tend to be slower than those for non-UI classes of similar complexity. This is due to the relative slowness of drawing components on-screen, and also because of pauses between keystrokes in these tests. Overall, the unit tests take a couple of hours to run on a very fast eight core machine with 16 gigabytes of memory.

Next are the **function tests**. Each functional requirement in our software specification leads to a test script in our testing specification. The function tests are Java classes that implement these test scripts. Each test takes about a minute to run, and there are about four hundred of them. The overall time to test these is about three and a half hours.

Third in line are the **load tests**. These stress and scalability tests take a long time to run—about two hours for the fifty-odd tests we have.

Sometimes it's not a clear-cut decision at which level to put a test. For example, our main server component, `RippledownServerImpl`, previously experienced problems that could only be replicated by a test that took more than an hour. In theory, this test should be at the unit level, since it is detecting a specific bug in a single class, and was originally written as such. However, as we need to be able to run our unit tests in a reasonable amount of time, we have refactored it as a load test.

What Language Should Our Tests Be In?

All tests should be in the primary development language of the project, namely Java, for the kinds of projects we are concerned with in this book. It is quite common to write tests in Python or some other scripting language, but we have found this to be unsatisfactory for a number of reasons.

For a start, our test code needs to be at least as good as our production code. It is easy for tests in a scripting language to be dismissed as mere 'test scripts'. If Python (say) is good enough for our tests, why is it not good enough for our production code?

Next, if we're writing a Java project, we can be sure that all team members are, or are becoming, proficient with Java. To require all developers to be as good at a second language as they are at Java is an unnecessary requirement.

Lastly, and perhaps most importantly, our test code needs to be maintained alongside our production code. For example, we need to have the confidence to be able to rename a method, or change its signature in our production code without breaking any test code. That is, we need to apply to our test code the same productivity and refactoring tools that are available in our primary language.

Is it Really Possible?

This talk about testing everything is all very well, but just how practical is it? Can we really write automated tests for our user interface components and other areas traditionally regarded as being too hard to test? Do we really have time to write all these unit and function tests? Given a pre-existing, largely untested codebase, isn't it so hard to move towards a fully-tested system, that it's not even worth trying?

A large part of this book is devoted to showing that user interfaces really can be tested, and that this is not so hard to do. We have done it with the LabWizard product suite and the Ikon Do It application that comes with the source code to this book. These have lots of automated user interface tests, proving that it can be done.

In answer to the question "Do we have time to write all these (expletive deleted) tests?", our experience is that we don't have time not to. It is great being able to sit with a customer, showing them new functionality, knowing that the new features have been thoroughly tested and will just work. Similarly, it's great when your customer has the expectation that each upgrade of your mission critical software at their site will be a painless process—in stark contrast to their experience of other software products! Having experienced this, one could never go back to a less "extreme" test methodology.

When faced with a pre-existing codebase that is poorly tested, we can feel overwhelmed by the amount of work to be done. In this situation, we need to remind ourselves that every single test we write is improving the situation, whereas any bit of untested code we add is making it worse. To be sure, getting to the 'Nirvana' of everything being tested can be a long hard journey. It certainly was for us with LabWizard:

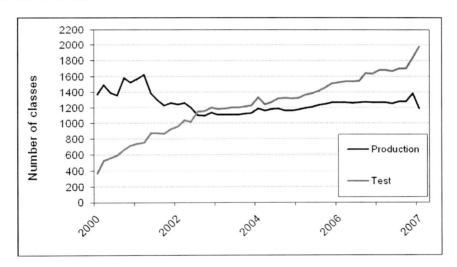

Over the period of this graph, about 60 releases of the software were made. Each release added many new features and bug fixes to the product suite. Yet because of the continual refactoring that is possible with a well-tested codebase, the total number of production classes has actually dropped, whilst the number of test classes has increased five-fold.

So yes, it is possible to follow the Extreme Testing approach, and it is so worthwhile that it just has to be done.

Summary

In this chapter, we have seen that all public classes need unit tests, and these tests should be written ahead of the classes they test, as far as possible. Each functional requirement in our software specification should correspond to a test class. All tests should be written in the main development language for our project and should be run as a part of our continuous build and integration process.

2
Basics of Unit Testing

In this chapter, we will write and test a simple class. This will demonstrate how the "test first" approach introduced in Chapter 1 can be applied in practice. We'll also look at the use of "Design by Contract", and the form that unit tests take.

A Simple Example

Suppose we are implementing a collection class `Relation` representing a mathematical relation. A `Relation` of size N simply consists of a set of pairs of objects (a_i, b_i), where i = 1, 2,..., N. The set of objects $\{a_i\}$ is called the domain, and the set of objects $\{b_i\}$ is called the range.

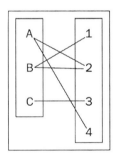

A relation between the letters {A, B, C} and the numbers {1, 2, 3, 4} is shown in the preceding figure. The letters form the domain and the numbers form the range. This relation contains the pairs (A, 2), (A, 4), (B, 1), (B, 2), and (C, 3).

A database table with two fields is a common example of a (persistent) `Relation`, with one field representing the domain, and the other representing the range.

Basics of Unit Testing

A `Relation` is similar to a `Map`, except that a given domain object can be related to more than one range object, whereas in a `Map` there can only be one range object for any given domain object.

A few of the core methods of `Relation` will be:

- Adding a pair of objects.
- Removing a pair of objects.
- Retrieving the size of the `Relation`.
- Checking if a given object is in the domain of the `Relation`.
- Retrieving the set of objects corresponding to the range of a given domain object.

How should we approach the implementation and testing of these methods?

The Default Implementation

According to the guideline introduced in Chapter 1, we define as few methods as we can get away with at first and give these methods any simple implementation that allows us to compile our class:

```
public class Relation<E, F> {
    /**
     * Inserts the specified pair.
     *
     * @pre domainElement not null
     * @pre rangeElement  not null
     */
    public void add(E domainElement, F rangeElement) {
    }
    /**
     * Removes the specified pair.
     *
     * @pre domainElement not null
     * @pre rangeElement  not null
     */
    public void remove(E domainElement, F rangeElement) {
    }
    /**
     * The number of (domainElement, rangeElement)
     * pairs in this relation.
     */
    public int size() {
        return 0;
    }
    /**
     * The set of domain elements of this relation.
     */
```

```
        public Set<E> domain() {
            return null;
        }
        /**
         * The set of range elements of this relation.
         */
        public Set<F> range() {
            return rangeImpl(null);
        }
        /**
         * True if and only if the given object
         * is in the range of this relation.
         */
        public boolean inRange(F obj) {
            return false;
        }
        /**
         * The range of the relation corresponding
         * to the specified domain element.
         */
        public Set<F> rangeFor(E domainElement) {
            return null;
        }
    }
```

Note that in a Map, the return value of rangeFor() would be an F, not a Set<F>. A more feature-rich Relation class would also have a method domainFor(), but this is not necessary here. A full implementation of Relation is provided in http://jpaul.sourceforge.net/javadoc/jpaul/DataStructs/Relation.html.

Test Cases

Each public method of Relation needs to be tested. However, for this example, we will just concentrate on add(). What sorts of test cases are likely to find errors in its implementation? The following are just a few examples:

- After calling add() for a pair (domainElement, rangeElement) not previously in the Relation, does the size of the Relation increase by one?
- After calling add(), does domainElement appear in the domain and rangeElement appear in the range?
- If we add the same pair a second time, does the size remain the same?
- If we add a pair where the domainElement is already in the domain, but the rangeElement is not in the range, does the size of the domain stay constant? Does the size of the Relation still increase by one? Does the size of the range increase by one?

- Similarly, if we add a pair where the `rangeElement` is already in the range, but the `domainElement` is not in the domain, does the size of the range stay constant? Does the size of the `Relation` still increase by one? Does the size of the domain increase by one?
- If we add a pair `(a, b)` where each of `a` and `b` is already in the `Relation` domain and range respectively, and the pair itself is new to the `Relation`, does the size of the `Relation` increase by one, and are the sizes of the domain and range unchanged?

One aspect that we don't have to test for is what happens if either `domainElement` or `rangeElement` is null. This is explicitly forbidden by the pre-condition comments:

```
* @pre domainElement not null
* @pre rangeElement  not null
```

The use of preconditions is part of a design methodology called "Design by Contract". This is a large and important topic in itself, but is directly relevant to unit testing, and so we give a brief overview here. See Meyer, *Object-Oriented Software Construction*, Second Edition (Prentice Hall, 2000; ISBN 0136291554) for a complete reference.

Design by Contract and Non-Defensive Programming

Pre-conditions are part of a contract between the user and implementer of a method. They specify the conditions that must be satisfied before a user is allowed to call the method. These may be conditions on the input parameters or conditions on the state of the object itself. By 'user' we mean the programmer writing the code that calls the method, not the end-user of the software.

The specific pre-conditions in the example above indicate that, before calling the `add()` method, it is the user's responsibility to ensure that the two parameters, `domainElement` and `rangeElement`, are not null.

Conversely, post-conditions specify the other half of the contract, namely, the responsibility of the implementer of the method with regards to any returned value and the subsequent state of the object.

So, from the implementer's point of view, the full contract is something like this: "If you, the caller, agree to your part of the contract (that is, satisfy my pre-conditions), then I the implementer will agree to my part of the contract (that is, satisfy my post-conditions)".

The use of pre-conditions in particular has the following main benefits:

- It simplifies the implementation of the method, as many (and possibly all) defensive checks can be avoided, such as checking for `null` parameters in the `add()` method above.
- As the method is guaranteed not to be called with unexpected inputs, the design of the return type is simpler. This is because there is no need to include some sort of error code in the return type, or generate an exception, to notify the user of this unexpected situation.
- The design of the test cases is simpler, as the tester can assume that the input parameters satisfy the pre-conditions, hence eliminating a possibly large number of test cases.
- The avoidance of defensive checks means that the execution speed of the method will be faster.
- The designer of the caller method has a clearer picture of the requirements.

For these reasons, we strongly recommend the following guideline:

Extreme Testing Guideline: We should document the pre-conditions for each public and protected method in our production classes, even if our development environment does not allow them to be checked at runtime.

Whilst other aspects of Design by Contract, such as post-conditions and invariants, are also recommended, if they can be applied, the use of pre-conditions alone probably gives the best "bang for the testing buck" due to the reduction in defensive code that we have to implement in our production classes and the corresponding reduction in the number of test cases that we have to write that check for illegal inputs.

Java does not natively support Design by Contract. We live in the hope that this will eventually be rectified. Apparently, the first specifications for Java did include support for Design by Contract, but it was dropped in order to meet a deadline. See `http://www.javaworld.com/javaworld/jw-12-2001/jw-1214-assert.html?page=5`. However, Java does have a simple assertion facility that can be used to check pre-conditions. When we implement our `add()` method, the first two lines will be:

```
assert domainElement != null;
assert rangeElement != null;
```

This goes a fair way towards implementing our pre-conditions, but is deficient for two reasons. The first problem is that the code executing the pre-conditions (the `assert` statements) is separate from the pre-conditions themselves. Only our vigilance will ensure that the comments accurately reflect the checks that are being done.

The second issue is that when we override a method, any new pre-conditions for the method in the subclass must be OR-ed with those in the superclass (and the post-conditions must be AND-ed, for reasons explained in Chapter 7). If our overriding method includes a call to the base method, any new pre-conditions will effectively be AND-ed with those from the superclass, which is exactly wrong.

Those really keen on using Design by Contract in Java may be interested in a tool called iContract, which uses a pre-processor to instrument code with contract checks that are written in the Javadoc. The tag `@pre`, which we used earlier, is the iContract notation for a pre-condition. We made a lot of use of iContract in implementing our core classes for LabWizard, and it was enormously valuable in helping us write well thought-out and error-free code. For example, one of our data structures is a `Tree` of indexed `Nodes` that contain `Rules`. We originally defined the `Tree` interface using contracts. Here are the contracts for the method by which a new leaf is added to the tree:

```
public interface Tree {
    /**
     * Insert the given node under the given parent node.
     * <br>
     *
```

```
     * @param node the node to be inserted
     * @param parent the node under which
     * to insert <code>node</code>
     *
     * @pre !nodes().contains( node )
     *          //<code>node</code>
     *          //is not already in the tree
     * @pre node.isLeaf()
     *          //<code>node</code> is a leaf
     * @pre node.parent() == null
     *          //<code>node</code> does not
     *          //yet have its parent set
     * @pre nodes().contains( parent )
     *          //the parent-to-be is in this tree
     * @pre node.intIndex() > parent.intIndex()
     *          //<code>node</code> has a
     *          //higher index than its parent-to-be
     * @pre (!parent.isLeaf()) implies
     *       (((Node) parent.children().last()).intIndex()
     *                      < node.intIndex())
     *          //if <code>parent</code> has any other
     *          //children, then the index of its youngest
     *          //child is less than the
     *          //index of <code>node</code>
     *
     * @post parent.children().last().equals( node )
     *          //<code>node</code> is the youngest
     *          //child of <code>parent</code>
     * @post node.isLeaf()
     *          //<code>node</code> is still a leaf
     * @post node.parent().equals( parent )
     *          //the parent of <code>node</code>
     *          //really is <code>parent</code>
     */
    public void addLeaf( Node node, Node parent );
    ....

}
```

iContract worked by running a pre-processor over the `Tree` interface and its implementations. This would produce java files that included the pre- and post-conditions as assertions at the start and end of methods. These "instrumented" java files would then be compiled in the regular way to produce class files instrumented with the pre- and post-conditions.

Basics of Unit Testing

We would run our unit tests against the instrumented classes. Due to the pre-conditions, our unit tests did not need to include examples where the node being added was null, was already in the tree, was not a leaf, or was not indexed correctly. In fact, any of these situations would have caused a pre-condition violation and test failure.

The post-conditions in the instrumented classes did a lot of the tedious checks that would have made testing a drag. Without post-conditions in our unit test for the addLeaf() method, we might add a Node to a tree and then check that:

- The Node is in the tree.
- The Node is the last child of its parent in the tree.
- The Node is still a leaf.
- The Node has the correct parent.

But all these assertions are actually in the post-conditions. Therefore, by simply adding a leaf to a tree, these checks would be made automatically in the instrumented classes.

Unfortunately we are no longer using the tool because of some annoying bugs and difficulties in integrating it into our automated build and test environment, but would love to get it going again. The original version of iContract has disappeared, but there is a new development of iContract, which can be found at http://icontract2.org/.

The use of pre-conditions is the antithesis of defensive programming. There are of course situations in which we need to write defensive code checks, for example when we are processing user input in a Swing application, or in a Web Service that millions of programmers have access to. Once we're "inside the fortress" of our own team's code though, it is far better to be offensive rather than be defensive, and this has always been the Java way: Array index checking is an example of it, as is the throwing of ConcurrentModificationExceptions by iterators in the collection classes. This can be summarized in yet another guideline:

[**Extreme Testing Guideline**: We should never defend against our own code.]

Attention to this guideline greatly simplifies our task as testers, as the production classes are not burdened with the additional complexity of having to be defensive. There is simply less functionality that we have to test.

There is one more issue about defensive coding that arises from this last point. Defensive coding often involves trying to work around situations in which there is actually no sensible course of action to take. For example, in our `Relation` implementation, what could we possibly try to do in the `add()` method for a pair in which either the domain or range is null? Or what could we do in a square root function when presented with a negative number? Attempting to code around these situations can be like having a discussion about the sound from one hand clapping. It's best not to be drawn into it.

Not surprisingly, Design by Contract is facilitated in an XP development where the team has "ownership" over the code, rather than individual developers having ownership over parts of the code. This means that each implementer must take responsibility for knowing how his or her method will be called. They must not take the easy way out by saying, "I don't know how Joe will call my method — that's his job — so I'll just defend against everything".

Test Code Example

Let's return now to our test class `RelationTest`. We know that we will be writing code to create a relation in a known state, many times. So we might as well write a convenience method for this straight away:

```
private Relation<String, String> r() {
    return new Relation<String, String>();
}
```

Convenience methods like this one take a lot of the pain out of writing tests.

Here is how we implement the test cases above for the `add()` method. The following is the implementation of our first batch of test cases for the add method.

```
public boolean addTest() {
    //a. After an add() of a pair (domainElement, rangeElement),
    //not previously in the relation, does the size
    //of the relation increase by one?
    Relation<String, String> relation = r();
    assert relation.size() == 0;
    relation.add( "a", "b" );
    assert relation.size() == 1;
    //b. After an add(), does domainElement appear in the
    //domain, and rangeElement appear in the range?
    relation = r();
    assert !relation.domain().contains( "a" );
```

```
assert !relation.inRange( "b" );
relation.add( "a", "b" );
assert relation.domain().contains( "a" );
assert relation.inRange( "b" );
//c. If we add the same pair in a second time,
//does the size remain the same?
relation = r();
relation.add( "a", "b" );
assert relation.size() == 1;
relation.add( "a", "b" );
assert relation.size() == 1;
//d. If we add in a pair where the domainElement
//is already in the domain, but the rangeElement
//is not in the range, does the size of the
//domain stay constant?
//Does the size of the Relation still increase by one?
//Does the size of the range increase by one?
relation = r();
assert relation.domain().size() == 0;
relation.add( "a", "b" );
assert relation.domain().size() == 1;
assert relation.size() == 1;
assert relation.range().size() == 1;
relation.add( "a", "c" );
assert relation.domain().size() == 1;
assert relation.size() == 2;
assert relation.range().size() == 2;
//e. If we add in a pair where the rangeElement
//is already in the range, but the domainElement
//is not in the domain,
//does the size of the range stay constant?
//Does the size of the relation still increase by one?
//Does the size of the domain increase by one?
relation = r();
relation.add( "a", "b" );
assert relation.domain().size() == 1;
assert relation.size() == 1;
assert relation.range().size() == 1;
relation.add( "x", "b" );
assert relation.domain().size() == 2;
assert relation.size() == 2;
assert relation.range().size() == 1;
```

```
    //f. If we add a pair (a, b) where each of a and b are
    //already in the relation domain and range respectively,
    //but the pair itself is new to the relation,
    //does the size of the relation increase by one,
    //and are the sizes of the domain and range unchanged?
    relation = r();
    relation.add( "a", "b" );
    relation.add( "c", "d" );
    assert relation.domain().size() == 2;
    assert relation.size() == 2;
    assert relation.range().size() == 2;
    relation.add( "a", "d" );
    assert relation.size() == 3;
    assert relation.domain().size() == 2;
    assert relation.range().size() == 2;
    relation.add( "c", "b" );
    assert relation.size() == 4;
    assert relation.domain().size() == 2;
    assert relation.range().size() == 2;
    return true;
}
```

Note that the test for the `add()` method is called `addTest()`. This convention will enable our test harness, GrandTestAuto, to check whether all required methods of the production class have in fact been tested.

Another thing to note is that when the test finds an error, it fails with an assertion, rather than by just returning `false`. Our experience in this matter is as follows.

Suppose that, a test uses some resource that needs to be cleaned up after the test is complete:

```
public boolean cancelExportTest() throws IOException {
    init();
    ui.createNewIkon( "test", 16, 16 );
       test body
    cleanup();
    return true;//If we get here, the test has passed.
}
```

(This example is from the 'Ikon Do It' unit tests.) If such a test fails, then the `cleanup()` call will be missed, which may cause subsequent tests to fail. So for tests of this nature, we might not want to use assert to do our checks.

On the other hand, the stack trace that accompanies an assertion failure indicates its location to the line, which is really useful for fixing the problem, especially if it is the kind of intermittent test failure we deal with in Chapter 20.

So we have to strike a balance between identifying failures precisely, and possibly causing a cascade of errors from a single test failure. On the whole we tend to use assertions a lot and only rarely have tests that actually return `false` to indicate failure, but this is not something we feel too strongly about.

Bootstrapping Our Implementation

Our test for `add()` makes use of a few other production class methods, such as `size()`, `domain()`, and so on. This is often unavoidable. Hence, we will need to provide non-trivial implementations of these in the production class, before the test for `add()` can be executed. That is, we will have to do a certain amount of bootstrapping. This is in effect swapping between the implementations of `RelationTest` and `Relation` before the full test and production class are completed.

In practice, this bootstrapping is not a problem. Our approach should be as follows:

[**Extreme Testing Guideline**: Implement and test the simpler, accessor methods of the production class first.]

Once the basic accessor methods are implemented, the more complex setter methods, that is, the ones that change the state of the object, such as `add()`, can then be tested. This is similar to the 'code a little, test a little' philosophy of JUnit, (which is expounded in http://junit.sourceforge.net/doc/testinfected/testing.htm) except that it is the other way around: test a little, code a little.

One reason we prefer to write the test for a method before the method itself is that writing the implementation tends to limit our approach to testing. The most effective tests come about when we really try hard to break the code we're testing. But nobody wants to break something that they have just created. If we write the tests first though, then it is as if we are creating a challenge for our method implementation to overcome, and we tend to feel happier about working this way.

Load Testing

One aspect of our class that we have not tested is its performance. We might be tempted to add tests such as:

- How long does it take to add several thousand pairs to the relation?
- How does the time taken to add a pair change, as the size of the relation changes?
- What is the memory footprint for a relation with a thousand pairs?

These load tests should not be included in the unit tests. The unit tests are to ensure correctness and must execute relatively quickly. So, as discussed in Chapter 1, it is better to put load tests in a separate testing pass. Of course, if the `Relations` that we build in our code only contain a dozen or so pairs, load testing the class at all would be a waste of time anyway. We will look more closely at load testing later, and one of the points we'll be making is that it can be a big waste of resources, and can harm a program, to optimize it too early in the development lifecycle.

Summary

In this chapter we have seen an example of how to write a unit test, and that our ideal of writing the tests for a class before the class itself will normally involve some degree of bootstrapping. We have seen that the use of the Design by Contract methodology to raise assertions for pre-condition failures greatly simplifies our code. We've also discussed the merits of having tests fail with an error, as opposed to simply returning a `boolean` value.

3
Infrastructure for Testing

Almost any project will have to deal with three test infrastructure issues, which are:

- Where should the test classes go?
- How should test data be managed?
- How should we deal with temporary files created by our tests?

In this chapter, we present solutions to each of these problems. We don't claim that these are the only good ways of setting up test infrastructure, simply because we don't think that there are single best approaches to these issues. What we present are solutions that have worked very well for us and can easily be tailored to the requirements and culture of other projects in other organizations.

Where Should the Unit Tests Go?

Three things we ask of our test classes are:

- It must be easy to check whether each production class that needs a test has a test.
- Test classes should not be in the same packages as our production classes.
- We need to be able to exclude our test classes from any deployed software.

The second of these requirements rules out the approach, seen in many older books, of putting test classes into the classes they are testing. Our test classes tend to be as complex as the classes they test, often having their own inner classes and even embedded test data. Cluttering up our production classes with unit tests is simply not necessary, given our goals of testing public and protected methods, as discussed in Chapter 1.

There are several reasons for excluding test classes from deployed software. The first is that we want to decrease the size of our deployed jars. However, a more fundamental reason is that we only want to ship tested code. It is essential that we can guarantee that no test code has leaked into our production classes. Without a clear separation at the package level of test code from production code, this can be a surprisingly difficult thing to ensure, particularly if a lot of test helper classes have been developed. It is all too easy for the test helper classes to 'help' also in the production code. For this reason, all of our test classes are in packages that end in ".test", ".functiontest", or ".loadtest", and none of our production code is in such packages. As we discussed in Chapter 1, there is no compelling need for explicit unit tests of package-private classes and methods. So this approach has worked very well for us.

Granted then that our test classes will be separate from the production classes, there are two obvious approaches for organizing test packages.

The first is to have a tree of test packages parallel to our production code.

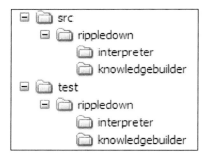

The second approach is to have the tests as a sub-package of the production package.

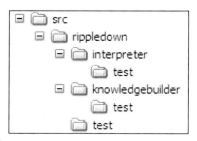

This is the approach we have taken. When we started the LabWizard code base, it was convenient to have the test packages in close proximity to their production packages. This is less of an issue with the improvements to development environments and other productivity tools that have taken place since then, and a lot of projects take the parallel hierarchies approach. It's not clear which method is actually better.

Where Should the Function and Load Tests Go?

As with the unit tests, function tests can be either in a parallel tree of packages or mixed in. We follow this second approach for the LabWizard product suite. The LabWizard (also known as "RippleDown") code base contains several applications: Interpreter, KnowledgeBuilder, Validator, and others. Each of these sub-applications has its own requirements specification and test specification. The documentation hierarchy is shown below:

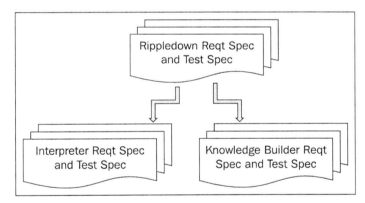

Each of the stand-alone products is associated with a group of packages. For example, the KnowledgeBuilder client application is mostly defined in `rippledown.knowledgebuilder.gui` and its sub-packages. The server-side components are in `rippledown.knowledgebuilder.core`. The function tests for the KnowledgeBuilder requirements are in `rippledown.knowledgebuilder.functiontest`, and similarly for the Interpreter and the Validator that we saw briefly in Chapter 1:

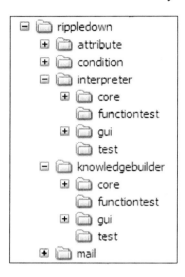

Our load tests are arranged similarly to the function tests in that the load tests for a major package are collected in a sub-package with the suffix ".loadtest". Keeping the load tests separate in this way makes it easy to run them in a separate testing pass.

Management of Test Data

Any software that communicates with other systems, or has a persistence layer, will need tests involving configured test data, and possibly very large amounts of it. For example, the LabWizard system has more than 3 Gigabytes of test data in about 1,000 files. This test data includes:

- **Messages that have been received from external systems**: A typical test of the input channel of an interface will read a configured test data file, parse it into some object, and then make assertions about the state of that object.
- **Messages that have been sent to external systems**: A typical test of the output channel of an interface will take some object in a known state, produce a message from it, and compare this with a configured message known to be correct.

- **Databases from client sites**: The LabWizard Knowledge Bases are persisted in B-Tree database files, copies of which are used in regression tests, or to reproduce bugs prior to implementing a fix. These databases take, by far, the greater portion of the space used in test data.

What Do We Require of a Test Data Management System?

The things we need from a system that manages our test data are:

- Test data must be held in a version control system, so that the version of test data we are using matches the version of production and test code.
- There must be an automated build step to retrieve the test data from the repository.
- The test data for a package must be readily identified. In particular, it must be easy to retrieve a particular test data file from within our tests.
- It must be easy to identify which test data files are no longer needed, so that our repository is not cluttered up with a lot of test data that is never used.

There are undoubtedly several good approaches to this issue. An example of the one we use is described next.

The LabWizard test data files are located in a directory structure that is parallel to the package structure for our source. That is, given a source directory tree of the form:

the test data files are located in the following directory tree.

Infrastructure for Testing

In the very early stages of the LabWizard development, test data was mixed in with the test packages themselves. This quickly became unworkable because of the sheer volume of test data that we had to manage. The main problem is that we often want to do a clean build that deletes all our source code and classes and compiles the project. If we've deleted three gigabytes of test data and need to retrieve it, such clean builds take too long. So, we have settled on storing the test data in a directory structure parallel to the code base.

With this test data arrangement, additional test classes can be automatically generated to make it easy to reference test data from within our tests, as follows:

A pre-processor, the `TestDataClassGenerator` program, is run over the test data directories. It creates a package hierarchy corresponding to the test data directory structure. The root of the package hierarchy can be chosen arbitrarily.

In the example above, the pre-processor would create the packages

`testdata.rippledown`,

`testdata.rippledown.attribute`,

`testdata.rippledown.condition`,

and so on. In each such package, the pre-processor creates a class, named according to the test data directory and containing public constants corresponding to the test data filenames in that test data directory.

For example, if there were test data files:

`/testdata/rippledown/interpreter/RuleTree1.xml`,

and

`/testdata/rippledown/interpreter/RuleTree2.xml`

then, the generated class `testdata.rippledown.interpreter.Interpreter` would contain the constants:

```
public static final String PATH = 
                    "...\\testdata\\rippledown\\interpreter\\";
public static final String RuleTree1_xml = PATH + "RuleTree1.xml";
public static final String RuleTree2_xml = PATH + "RuleTree2.xml";
```

Once the classes that the pre-processor has generated are compiled, our tests can access the test data files using the automatically generated constants rather than hard-coded filenames. For example, suppose the test class `rippledown.interpreter.test.RuleTreeTest` requires configured rule trees serialized in the files, `RuleTree1.xml` and `RuleTree2.xml`. We can retrieve this configured data from within the test as follows:

```
File configuredTree1 = new File( Interpreter.RuleTree1_xml );
File configuredTree2 = new File( Interpreter.RuleTree2_xml );
```

Test data is configured in the same repository as the source code. Our Ant task for a totally clean build executes the following steps:

- Delete all source code, test code, and test data from the test workstation.
- Retrieve the configured source code and test data from the repository.
- Compile all production classes and the pre-processor itself.
- Run the pre-processor over the test data directory.
- Compile all test classes.
- Run the tests.

Running the pre-processor to generate the test data constant classes takes about five seconds on a fairly low-powered machine.

Managing test data in this way meets all the objectives we listed earlier. Public constants that are not referenced in our source code can be easily identified by our Java development environment, thus identifying any test data files that are no longer used.

A version of `TestDataClassGenerator` is used in the software for this book, and the source code for it can be found in the package `jet.testtools`. The unit tests for this class are also worth looking at.

Temporary Files

If our software interacts with the file system of the computer on which it runs, then our unit tests will necessarily involve a lot of interaction with the file system too. This is a situation where some well-chosen convenience methods can really cut down the drudgery of writing tests.

We want the files created by our tests to be easily available for investigation, should a test fail. This implies two things. First of all, our temporary files directory should be in some convenient location. Secondly, the temporary files need to survive the JVM. This second requirement rules out the `createTempFile()` methods in `java.io.File`.

Our approach has been to have a test helper class `Files` that has the following methods

```
public static File tempDir();
public static void cleanDirectory( File dir );
public static File cleanedTempDir();
```

as well as several others for copying files and so on, which are discussed in more detail in Chapter 14. Our `Files` class is very simple and relies only on the basic Java libraries. For a more advanced helper class, we might enlist the aid of a library such as `org.apache.commons.io`. It's interesting to note that even for this dedicated IO toolset, the tests include a helper class that has a convenience method for getting the temporary directory—see `http://org.apache.commons.io.testtools.FileBasedTestCase` for details. (This very useful library and its tests can be downloaded from `commons.apache.org/downloads/download_io`.)

Our implementation of `tempDir()` is:

```
public static File tempDir() {
    String userDir = System.getProperty( "user.dir" );
    File result = new File( userDir, "temp" );
    result.mkdirs();
    return result;
}
```

This puts the temporary directory under the user directory, which can of course be set using the flag "`-Duser.dir=...`" in the JVM startup script. If setting the value of the "`user.dir`" variable was a problem, we could use our own parameter for this.

Summary

In this chapter, we have seen that unit tests are kept in `.test` sub-packages of the packages they test. This makes it easy to distinguish them from the application classes themselves. Function and load tests are in `.functiontest` and `.loadtest` packages respectively. We will return to this arrangement in Chapter 19.

The code for this book includes a tool, `TestDataClassGenerator`, that makes it easy to write convenience methods to handle configured test data files. This tool also helps us detect test data that is no longer being used so can be deleted from our source control repository.

Finally, we presented the easily-solved problem of where to put temporary files created during tests. Again, there is a class provided for this.

4
Cyborg—a Better Robot

Over the next seven chapters, we will be concentrating on techniques for testing user interfaces. Our most basic tool for this is `Cyborg`.

The `java.awt.Robot` class, introduced in JDK 1.3, allows us to generate native mouse and keyboard events in our test programs. This makes it possible for us to write automated tests for user interfaces.

Now, `Robot` is a low-level class, having methods such as `keyPress(int keycode)` and `mouseMove(int x, int y)`. In our tests, by contrast, we want to type long strings of text, click elements of lists, close windows, activate menus with mnemonics and accelerators, and do myriad other things. To bridge the gap, we have developed a wrapper class called `Cyborg`, which has convenience methods for doing the kinds of high-level operations we are interested in.

The `Cyborg` class presented here and in the source code is similar to code originally produced at Pacific Knowledge Systems by Martin Manttan and Van Hai Ho.

The Design of Cyborg

`Cyborg` uses `Robot` to do the actual work of generating events. We could have made `Cyborg` extend `Robot`, but have chosen not to, for two main reasons. The first is that `Robot` has a lot of low-level methods that will not be used, and which we don't want showing up in the auto-completion boxes of our development environment. The second is that the constructor of `Robot` has a `throws` clause that is quite irritating to deal with in unit tests. Neither of these reasons is particularly compelling. So if you'd prefer `Cyborg` to extend `Robot`, well, you've got access to the source code, but for us `Cyborg` starts like this:

```
public final class Cyborg {
    private Robot robot;
    public Cyborg() {
```

```
            try {
               robot = new Robot();
               robot.setAutoWaitForIdle( true );
            } catch (AWTException e) {
                e.printStackTrace();
                assert false : "Could not create robot";
            }
        }
    }
```

Note the call `robot.setAutoWaitForIdle(true)`, which causes the `Robot` to wait for each event to be consumed before generating the next. Tuning the behavior of the `Robot` in this way is essential to the working of `Cyborg`. If `Cyborg` extended `Robot`, clients would be free to undo this setting. This is a third reason why `Cyborg` does not extend `Robot`.

Using the Keyboard

By far, the most common use of `Cyborg` is to enter keyboard data. This can mean typing in text, but also includes operations such as using the *Delete* key, tabbing, using the arrow keys, and so on. It also includes using key combinations such as *Alt+F*, or *Ctrl+F*.

The most basic keyboard operation in `Cyborg` is to type a single key, as represented by a `KeyEvent` constant:

```
    /**
     * Press then release the given key. Assert fail if the key could
     * not be typed (which presumably means that the current platform
     * does not support that key).
     *
     * @param key a KeyEvent constant, eg KeyEvent.VK_0.
     */
    private void typeKey( final int key ) {
        try {
           robot.keyPress( key );
           pause();
           robot.keyRelease( key );
           pause();
        } catch (Exception e) {
            assert false : "Invalid key code. As int: '"
                            + key + "', as char: '" + (char) key + "'";
        }
    }
```

The `pause()` is a 20 millisecond sleep for the current thread. Through trial and error (lots of the latter), we've found this pause to be necessary for obtaining robust tests. It also seems appropriate to have some gap between operations that would be performed by a human. A downside of this is that it limits the speed at which our tests will run. This is mainly a problem in tests that type in significant amounts of text, but we work around this by putting long text into the Clipboard and pasting it instead of typing it using key events.

This basic operation already allows us to define a lot of convenience methods:

```
public void delete() {
    typeKey( KeyEvent.VK_DELETE );
}
public void escape() {
    typeKey( KeyEvent.VK_ESCAPE );
}
public void enter() {
    typeKey( KeyEvent.VK_ENTER );
}
```

There are a lot more such methods in `Cyborg`.

The second most basic keyboard operation is to type a key together with *Shift*, *Alt*, or *Control*. This is achieved by a key press for the modifier, typing the key, and then releasing the modifier:

```
private void modifiedKey( final int key, final int modifier ) {
    robot.keyPress( modifier );
    pause();
    typeKey( key );
    robot.keyRelease( modifier );
    pause();
}
```

This allows a new batch of convenience methods:

```
public void altF4() {
    modifiedKey( KeyEvent.VK_F4, KeyEvent.VK_ALT );
}
public void ctrlA() {
    modifiedKey( KeyEvent.VK_A, KeyEvent.VK_CONTROL );
}
```

It also allows us to type an arbitrary character:

```
private void type( char c ) {
    switch (c) {
            //Digits.
        case '0':
```

```
            typeKey( KeyEvent.VK_0 );
            break;
        ...//etc
        //Lower case letters.
        case 'a':
            typeKey( KeyEvent.VK_A );
            break;
        ...//etc
        //Upper case letters: use the shift key.
        case 'A':
            modifiedKey( KeyEvent.VK_A, KeyEvent.VK_SHIFT );
            break;
        ...//etc
        //Symbols that are shifted numbers.
        case '!':
            modifiedKey( KeyEvent.VK_1, KeyEvent.VK_SHIFT );
            break;
        ...//etc
        //Other symbols.
        case ' ':
            typeKey( KeyEvent.VK_SPACE );
        case '-':
            typeKey( KeyEvent.VK_MINUS );
            break;
        ...//etc
        default:
            assert false : "Did not know how to type: '" + c + "'";
    }
```

In the code listing above, there is a very long switch statement that could be replaced by some reflective code that looks up the relevant `KeyEvent` constant based on the character being typed, with a shift added for upper-case letters. However, we find the switch statement easier to follow.

From a single arbitrary character it is easy to type any `String`:

```
public void type( final String str ) {
    final char[] chars = str.toCharArray();
    for (int i = 0; i < chars.length; i++) {
        type( chars[i] );
    }
}
```

To enter a string quickly, we put it into the system clipboard and paste it:

```
public void enterText( final String str ) {
    if (str.length() < 3) {
        type( str );
    } else {
        putIntoClipboard( str );
        paste();
    }
}
/**
 * Puts the given string into the system clipboard.
 */
public void putIntoClipboard( final String str ) {
    Toolkit toolkit = Toolkit.getDefaultToolkit();
    Clipboard clipboard =
            toolkit.getSystemClipboard();
    StringSelection selection = new StringSelection( str );
    clipboard.setContents( selection, selection );
}
/**
 * Pastes whatever is in the system clipboard
 * using CTRL_V.
 */
public void paste() {
    modifiedKey( KeyEvent.VK_V, KeyEvent.VK_CONTROL );
}
```

Mousing Around

`Robot` provides us with methods for moving the mouse, pressing and releasing the buttons, and scrolling with the wheel. Typically, in a test, we want to click the mouse at a particular point. For example, at the position of an item in a list or table, right-click somewhere to bring up a contextual menu, double-click a component, or drag between two points. The class UI used in these code snippets is a toolset for interacting with Swing components in a thread-safe manner and is the subject of Chapters 8 and 10.

The most basic operation is a left mouse click:

```
/**
 * Presses and releases the left mouse button.
 */
public void mouseLeftClick(  ) {
    robot.mousePress( InputEvent.BUTTON1_MASK );
    robot.mouseRelease( InputEvent.BUTTON1_MASK );
}
```

Based on this, we can click at a point, and click a component:

```
/**
 * Moves the mouse to the given screen coordinates,
 * and then presses and releases the left mouse button.
 */
public void mouseLeftClick( Point where ) {
    robot.mouseMove( where.x, where.y );
    mouseLeftClick();
}
/**
 * Left-clicks the mouse just inside the given component.
 * Will not work for tiny components (width or
 * height < 3).
 */
public void mouseLeftClick( Component component ) {
    Point componentLocation = UI.getScreenLocation( component );
    Point justInside = new Point( componentLocation.x + 2,
                                  componentLocation.y + 2 );
    mouseLeftClick( justInside );
}
```

There are similar methods for right-clicking and double-clicking.

A slightly more challenging method to implement is dragging the mouse. Our first attempt at implementing this was as follows:

- Move the mouse to the start point.
- Press the left mouse button.
- Move the mouse to the end point.
- Release the left mouse button.

Unfortunately this doesn't work. When we use Robot to move the mouse to a point, it goes straight there, without passing through any of the points in between the original position and the destination. Maybe there's some kind of space-time wormhole in use that only Sun engineers know about!

Our implementation finds the list of all points between the start and end points of the drag and traverses them with the left mouse button down:

```
/**
 * Moves the mouse to <code>start</code>, presses
 * the left mouse button, moves the mouse to
 * <code>end</code> and releases the button.
```

```
    */
    public void mouseDrag( Point start, Point end ) {
        java.util.List<Point> points = lineJoining( start, end );
        robot.mouseMove( start.x, start.y );
        robot.mousePress( InputEvent.BUTTON1_MASK );
        for (Point o : points) {
            robot.mouseMove( o.x, o.y );
            pause();
        }
        robot.mouseRelease( InputEvent.BUTTON1_MASK );
    }
```

The (slightly) tricky part is to work out the list of points between the start and the end. Details of this can be seen in the source code.

Checking the Screen

Apart from the keyboard and mouse event generation capabilities that we've just seen, Robot also gives us access to the pixels on the screen via the methods getPixelColor() and createScreenCapture(). These allow us to test animations and other programs where we are tightly controlling the screen painting. Cyborg has a thread-safe wrapper for getPixelColor() and also a method that checks the color of a pixel and prints out a useful debugging message if it is not as expected:

```
    /**
     * Retrieves the color of the pixel at the given
     * co-ordinates; does this from within the event thread.
     */
    public Color safelyGetPixelColour( final int x, final int y ) {
        final Color[] result = new Color[1];
        UI.runInEventThread( new Runnable() {
            public void run() {
                result[0] = robot.getPixelColor( x, y );
            }
        } );
        return result[0];
    }
    /**
     * Check that the pixel at the given co-ordinates
     * of the given component is the expected color.
     * The co-ordinates are in the co-ordinate space
     * of the component; to check the top left pixel
     * use relativeX = 0, relativeY=0.
     */
```

```
public void checkPixel( final int relativeX,
                        final int relativeY,
                        final Color expected,
                        final Component component ) {
    int actualX = relativeX + component.getLocationOnScreen().x;
    int actualY = relativeY + component.getLocationOnScreen().y;
    Color actual = safelyGetPixelColour( actualX, actualY );
    String errMsg = "relativeX: " + relativeX + ", relativeY: "
              + relativeY + ", actualX: " + actualX
                                    + ", actualY: "
              + actualY + ", expected: " + expected
              + ", got: " + actual;
    assert actual.equals( expected ) : errMsg;
}
```

Summary

Cyborg provides a great many useful tools for automating the testing of Java applications. Here we have had a quick look at some of the features of Cyborg. More details are to be found in the source code and unit tests and, of course, in the later chapters. We will be using Cyborg a lot in Chapters 7, 8, 9, and 10.

5
Managing and Testing User Messages

In Java, we use resource bundles to store the text that is presented in our user interfaces. This has a number of benefits. For example, it is then very easy to produce localized versions of our software. In this chapter, we introduce a system of managing and testing resource bundles that achieves two important goals.

First, we will be able to thoroughly unit test our resource bundles. This eliminates a lot of mistakes that would otherwise be found only with laborious manual testing.

Second, we will ensure that keyboard shortcuts can be used to activate all of our components. This is an important aspect of usability, and keyboard shortcuts are the way we will be activating our components in automated user interface tests — this features heavily in the later chapters.

We begin with a discussion of Java's inbuilt support for localization.

Some Problems with Resource Bundles

Java's internationalization support means that with a small amount of discipline, it is possible to create applications that can be easily translated into different languages. The required discipline is that the text that is displayed to the users is never hard-coded into our software. Instead, our programs use keys to strings that are in language-specific resource bundles. At run time, the appropriate resource bundle is loaded into the system and the software presents messages in the language of the user.

A specific example might work as follows. Suppose we are defining a **File** menu. We will have a resource bundle that is loaded from a properties file:

```
resources = ResourceBundle.getBundle( "Messages" );
```

To label our **File** menu is simple:

```
fileMenu = new JMenu( resources.getString( "file" ) );
```

This works for a fairly small system, but has a number of problems.

First, there is no guarantee that the constant `"file"` is the name of a constant in our resource bundle. Mistakes like this are generally cleared up through free-play testing, but this is not guaranteed. For example, the resource may be in some seldom-used GUI component which free-play testing may easily miss. Wouldn't it be better if we could guarantee that these errors never occur?

The second problem is that as our user interface classes change, the properties file can accumulate unused keys. This is really just a nuisance, but it would be better if it were easy to eliminate such 'junk keys'.

A third problem is that just as a button will generally have a mnemonic, accelerator key, tool-tip, and icon, we often do not just want a single key, rather a whole family, for a concept such as "file". We want convenience methods for obtaining these resources. We also want to ensure that if there is a mnemonic, it is actually a character in the label, and that if there is an icon, it does really point to an image.

A final problem concerns formatted messages. Suppose that we want to present to the user a message such as "The user account 'Snoopy' does not have sufficient privileges for this operation." To produce this using resource bundles, we would typically proceed as follows:

```
String noAcctMsg = resources.getString( "low_privs" );
String toUser = MessageFormat.format( noAcctMsg, "Snoopy" );
```

The string `noAcctMsg` would have a placeholder `{0}`, which the Java `MessageFormat` class replaces with the value `Snoopy`. Now, suppose that we wish to make our message more helpful: "The user account 'Snoopy' does not have Administrator privileges." This is snappier and shows what privileges Snoopy needs. To do this, we change our value for the `low_privs` key and add another variable to our formatting call:

```
String noAcctMsg = resources.getString( "low_privs" );
String toUser = MessageFormat.format( noAcctMsg, "Snoopy",
                                      "Administrator" );
```

This will cause problems if the number of placeholders in `noAcctMsg` is not two. It is very easy for mistakes like these to creep into our code, and the earlier we find them, the better.

In this chapter, we describe a methodology for resource management that addresses all these issues, and at the same time makes it easier to write tests for our applications. Note that although we are presenting our resource management system in the context of Swing user interfaces, it could work equally well in any other situation where messages need to be translated, such as in a JavaServer Faces application.

A Solution

Our approach is to have a single class that manages the resource bundle for a package that requires internationalization. This class provides a method for getting formatted messages as well as methods for getting the labels, mnemonics, tool-tips, accelerators, and icons. The class extends `jet.util.UserStrings`, in which these methods are defined. The properties keys are defined as class constants. The final detail is that the class implements the singleton pattern. If this is a lot to absorb in one paragraph, fear not, we'll look at an example of a `UserStrings` subclass soon. For now, let's see how such classes are used in code where we are defining user text.

With our approach, a formatted message is obtained as follows. First, we get a hold of the singleton instance of the relevant `UserStrings` subclass:

```
UserStrings us = LoginUserStrings.instance();
```

Then, we apply the `message()` method using a class constant as the key:

```
String msg = us.message(
    LoginUserStrings.UNKNOWN_USER_MSG1, userName );
```

To get a button label, we call the `label()` method:

```
String label = us.label( LoginUserStrings.USER_NAME );
```

Three advantages of this approach are already apparent:

- The handling of the resource bundle is encapsulated in the `UserStrings` class, and we don't need to bother with it here.
- By using constants in the `LoginUserStrings` class as keys, we are eliminating errors due to misspelled keys.
- Again, because we're using constants, we can use code-completion in our Java development environment, a small but genuine productivity gain.

Here is how it works. Each resource in the properties file is either a message or part of a family of resources associated with a particular item. (Such a family might contain the label, mnemonic, and tool-tip for a button, for example.) The resource keys are split into two groups accordingly.

For a message resource, the key will end with the suffix _MSGn, where n is the number of formatting placeholders in the value. For example:

```
WRONG_PASSWORD_MSG0=Wrong password!
UNKNOWN_USER_MSG1=Unknown user: {0}
INSUFFICIENT_PRIVILEGES_MSG2={0} does not have {1} privileges!
```

For a family of resources, all the keys share a root value and have a suffix that denotes their use:

```
USER_NAME_LA=Name
USER_NAME_MN=n
USER_NAME_TT=User name - case insensitive.
```

The keys for messages are constants in the particular UserStrings subclass, as are the roots of the resource family keys:

```
public static final String
    INSUFFICIENT_PRIVILEGES_MSG2 = "INSUFFICIENT_PRIVILEGES_MSG2";

public static final String USER_NAME = "USER_NAME";
```

As we shall see, this design allows us to thoroughly test our resource bundles using Java's Reflection API.

A further gain is that by searching for unused class constants, we can identify unused resources in our properties files.

Before seeing how the tests will work, let's look in detail at our wrapper classes.

The UserStrings Class

The base class for all resource string wrapper classes is UserStrings:

```java
public class UserStrings {
    public static final String LABEL_EXT = "_LA";
    public static final String MNEMONIC_EXT = "_MN";
    public static final String ACCELERATOR_EXT = "_AC";
    public static final String TOOLTIP_EXT = "_TT";
    public static final String ICON_EXT = "_IC";
    private ResourceBundle rb;

    protected UserStrings() {
        try {
            rb = ResourceBundle.getBundle( getClass().getName() );
        } catch (Exception e) {
            e.printStackTrace();
```

```java
            assert false : "Could not load resource bundle.";
        }
    }
    /**
     * Formats the message for the given key with the specific
     * details provided.
     */
    public String message( String key, Object details ) {
        String value = rb.getString( key );
        assert key.endsWith( "" + details.length ) :
                "Mismatch between placeholders and variables. " +
                "Expected " + details.length +" places. Value is:
                                                        " + value;
        return MessageFormat.format( value, details );
    }
    public Integer mnemonic( final String key ) {
        String message = rb.getString( key + "_MN" ).toUpperCase();
        if (message.length() == 0) return null;
        String fieldName = "VK_" + message.charAt( 0 );
        try {
            return KeyEvent.class.getField( fieldName ).getInt( null );
        } catch (Exception e) {
            return null;
        }
    }
    public String toolTip( String key ) {
        String fullKey = key + TOOLTIP_EXT;
        if (!rb.containsKey( fullKey )) return null;
        return rb.getString( fullKey );
    }
    public String label( String key ) {
        return rb.getString( key + LABEL_EXT );
    }
    public KeyStroke accelerator( String key ) {
        String fullKey = key + ACCELERATOR_EXT;
        if (!rb.containsKey( fullKey )) return null;
        String message =  rb.getString( fullKey );
        return KeyStroke.getKeyStroke( message );
    }
    public Icon icon( String key ) {
        Icon icon = null;
        String fullKey = key + ICON_EXT;
```

```java
            if (!rb.containsKey( fullKey )) return null;
            String iconName = rb.getString( fullKey );
            if (iconName.length() > 0) {
                String packagePrefix = getClass().getPackage().getName();
                packagePrefix = packagePrefix.replace( '.', '/' );
                iconName = packagePrefix + "/" + iconName;
                icon = new ImageIcon(
                        ClassLoader.getSystemResource( iconName ) );
            }
            return icon;
        }
    }
```

A typical subclass consists mostly of constants:

```java
    public class LoginUserStrings extends UserStrings {
        //Constants for message keys.
        public static final String ADMINISTRATION_MSG0 =
                                        "ADMINISTRATION_MSG0";
        ...
        //Constants for resource family keys.
        public static final String USER_NAME = "USER_NAME";
        public static final String PASSWORD = "PASSWORD";
        ...
        //Static instance and private
        //constructor for the singleton pattern.
        private static LoginUserStrings instance;
        public synchronized static LoginUserStrings instance() {
            if (instance == null) {
                instance = new LoginUserStrings();
            }
            return instance;
        }
        private LoginUserStrings() {}
    }
```

The resource bundle for this class will be:

```
    ADMINISTRATION_MSG0=Administration
        ...
    USER_NAME_LA=Name
    USER_NAME_MN=n
    USER_NAME_TT= User name - case insensitive.
```

```
PASSWORD_LA=Password
PASSWORD_MN=p
PASSWORD_TT=Password - ask your administrator if you have forgotten
                      it.
...
```

Here is a typical snippet of code that uses `LoginUserStrings` as an instance variable `us`:

```
JLabel userNameLabel =
            new JLabel( us.label( LoginUserStrings.USER_NAME ) );
userNameLabel.setDisplayedMnemonic(
                    us.mnemonic( LoginUserStrings.USER_NAME ) );
```

ResourcesTester

We have written a class, `jet.util.ResourcesTester`, which allows us to unit-test our `UserStrings` subclasses and their associated resource bundles. For a `UserStrings` subclass and its properties file, `ResourcesTester` will check that:

- All class constants correspond to resources.
- All resources keys are represented by constants.
- The values for message constants actually have the number of placeholders implied by their name.
- The resource family for a constant is well-defined in that:
    - The mnemonic for a label is a character in the label.
    - The accelerator (if it exists) really is a key stroke.
    - The icon location (again, if not blank) actually points to an image file.

`ResourcesTester` is very easy to use. Here is the unit test for `LoginUserStrings` and its resource bundle:

```
public class LoginUserStringsTest {
    public boolean instanceTest() throws Exception {
        //The next line throws an exception if there's a problem.
        ResourcesTester.testAll( LoginUserStrings.instance() );
        return true;
    }
}
```

How ResourcesTester Works

The implementation of `ResourcesTester` is a good use of Reflection, as we'll now see. Those readers not so interested in these technicalities can skip ahead to the next section.

In order to test the resources and constants in a `UserStrings` subclass, `ResourcesTester` needs both the resource bundle and its associated wrapper class. This data is obtained with a `UserStrings` instance:

```
public class ResourcesTester {
    //The resource bundle for the UserStrings subclass.
    private ResourceBundle rb;

    //Instance of the class under test.
    private UserStrings us;

    ...
}
```

In order to run checks on the class constants and their values, it is convenient to have these in a collection:

```
//Map of the constant strings in the class under test
//to their values. We need to look at field names and values,
//this map makes it easy to do so.
private Map<Field, String> constants =
                        new HashMap<Field, String>();
```

One other thing that is needed is a regular expression for recognizing message constants:

```
//Recognizes constants that are for messages, and extracts
//from the constants the implied number of placeholders.
private Pattern messagesPattern =
                        Pattern.compile( ".*_MSG([0-9]*)" );
```

The constructor is private and mostly involves using Reflection to load the map of constants to their values:

```
private ResourcesTester( UserStrings us ) {
    this.us = us;
    rb = ResourceBundle.getBundle( us.getClass().getName() );

    //Load the key constants.
    Field[] fields = us.getClass().getDeclaredFields();
    for (Field field : fields) {
        int m = field.getModifiers();
        if (Modifier.isPublic( m ) && Modifier.isStatic( m ) &&
```

```
                Modifier.isFinal( m )) {
            try {
                constants.put(
                        field, (String) field.get( null ) );
            } catch (IllegalAccessException e) {
                e.printStackTrace();
                assert false : "Could not get field value.";
            }
        }
    }
}
```

The only public method in `ResourcesTester` creates an instance and does a two-stage test:

```
public static void testAll( UserStrings us ) {
    ResourcesTester tester = new ResourcesTester( us );
    tester.checkClassConstants();
    tester.checkResourceKeys();
}
```

In the first stage of the test, we iterate over the class constants and run checks according to whether they are a message constant or a resource family constant:

```
private void checkClassConstants() {
    for (Field constant : constants.keySet()) {
        String value = constants.get( constant );
        Matcher matcher = messagesPattern.matcher( value );
        if (matcher.matches()) {
            checkMessageConstant( matcher, constant );
        } else {
            checkResourceFamilyConstant( constant );
        }
    }
}
```

The actual checking of message constants involves determining from the suffix of the constant how many placeholders the associated resource value should have, and then checking that precisely that many slots exist:

```
private void checkMessageConstant( Matcher m, Field constant ) {
    //First check that the name is in fact the value.
    String constantValue = constants.get( constant );
    String fieldName = constant.getName();
    assert constantValue.equals( fieldName ) :
```

```
                "The field " + constant + "did not have " +
                            "its name the same as its value.";
        String numberAtEnd = m.group( 1 );
        int numSlots = Integer.parseInt( numberAtEnd );
        String value = rb.getString( constantValue );
        if (numSlots == 0) {
            //Check that there are in fact no placeholders.
            assert!value.contains( "ZZZZ" );//Sanity check
            String formatted = MessageFormat.format( value, "ZZZZ" );
            //If ZZZZ is in the formatted string, there was one.
            assert!formatted.contains( "ZZZZ" ) :
                        "Should be no placeholders for: " + constant;
        } else {
            //Check that it has sufficient parameter slots.
            //Do this by formatting it with some strange values.
            Object[] params = new String[numSlots];
            for (int i = 0; i < params.length; i++) {
                params[i] = "ZZZZ" + i + "YYYY";
            }
            String formatted = MessageFormat.format( value, params );
            //All of these strange values should
            //appear in the formatted string.
            for (Object param : params) {
                assert formatted.contains( param.toString() ) :
                            "Missing parameter slot for: " + constant;
            }
            //There should be no placeholders left over. Format the
            //formatted string, with a new strange value, to check this.
            formatted = MessageFormat.format( formatted, "WWWWQQQQ" );
            assert!formatted.contains( "WWWWQQQQ" ) :
                        "Extra parameter slot for: " + constant;
        }
    }
}
```

Testing resource family constants involves checking the consistency of the set of resources. If the mnemonic is defined, it should be a character in the label; if the accelerator is defined, it really should represent a keystroke; and if the icon is defined, it really should represent an image.

```
    private void checkResourceFamilyConstant( Field constant ) {
        String value = constants.get( constant );

        //If the accelerator string is defined, then it must
        //represent a valid accelerator.
        try {
```

```java
            String ac = rb.getString(
                        value + UserStrings.ACCELERATOR_EXT );
            if (ac != null && !ac.equals( "" )) {
                assert us.accelerator( value ) != null :
                                    ac + " is not a KeyStroke";
            }
        } catch (MissingResourceException e) {
            //The accelerator is not defined.
        }

        //Check that if the mnemonic and label exist,
        //then the mnemonic is a character in the label.
        try {
            String mn = rb.getString( value
                            + UserStrings.MNEMONIC_EXT );
            String label = us.label( value );
            if (mn != null && mn.length() > 0 && label != null) {
                if (label.toLowerCase().indexOf(
                                    mn.toLowerCase() ) < 0) {
                    String errorMessage = "Mn not in label. Key: '"+
                            "', label text: '" + label + "', mnemonic: '"
                            + mn + "'";
                    assert false : errorMessage;
                }
            }
        } catch (MissingResourceException e) {
            //Label or mnemonic not defined, so nothing to check.
        }

        //Check that if an icon is defined, it actually represents
        //an image resource.
        try {
            String iconVal = rb.getString( value
                                + UserStrings.ICON_EXT );
            if (iconVal != null && iconVal.length() > 0) {
                assert us.icon( value ) != null : "No icon: " + iconVal;
            }
        } catch (MissingResourceException e) {
            //Icon not defined, so nothing to check.
        }
    }
```

At this point, the `ResourcesTester` has checked that all class constants represent messages or resource families, and that all messages and resource families represented by class constants are correct. The second part of the test is to check whether all resource keys are represented by constants. This is easily done by using the pattern matcher:

```
private void checkResourceKeys() {
    for (String keyStr : rb.keySet()) {
        Matcher matcher = messagesPattern.matcher( keyStr );
        if (matcher.matches()) {
            //It should be a constant name.
            assert constants.values().contains( keyStr )
                            : "Not a constant: " + keyStr;
        } else {
            //It should be a member of the resource family for
            //for a constant.
            boolean endingOk = (
                keyStr.endsWith( UserStrings.ACCELERATOR_EXT ) ||
                keyStr.endsWith( UserStrings.ICON_EXT ) ||
                keyStr.endsWith( UserStrings.LABEL_EXT ) ||
                keyStr.endsWith( UserStrings.MNEMONIC_EXT ) ||
                keyStr.endsWith( UserStrings.TOOLTIP_EXT ));
            assert endingOk : "Unknown resource type: " + keyStr;

            //The base of the key should be a constant.
            //All _EXT constants have length 3.
            int limit = keyStr.length() - 3;
            String keyBase = keyStr.substring( 0, limit );
            assert constants.values().contains( keyBase ) :
                        "No constant for: '" + keyBase + "'";
        }
    }
}
```

Getting More from UserStrings

`UserStrings` can be made even more useful with the addition of factory methods for `JButton`, `JMenuItem`, and so on. For example:

```
public JButton createJButton( Action a, String key ) {
    JButton result = new JButton( a );
    result.setText( label( key ) );
    Integer mnemonic = mnemonic(key);
    if (mnemonic != null) {
        result.setMnemonic( mnemonic );
```

```
        }
        String toolTip = toolTip( key );
        if (toolTip.length() > 0) {
            result.setToolTipText( toolTip );
        }
        Icon icon = icon( key );
        if (icon != null) {
            result.setIcon( icon );
        }
        KeyStroke accelerator = accelerator( key );
        if (accelerator != null) {
            a.putValue( Action.ACCELERATOR_KEY, accelerator );
        }
        result.setName( key );
        return result;
    }
```

Note that, as with the other methods in `UserStrings`, we always call this method with the key a class constant. For example:

```
cancelButton = us.createJButton( cancel, LoginUserStrings.CANCEL );
```

If we spell CANCEL wrong, we will get a compiler error.

The second to last line of this method sets the name of the button. This allows us to get a reference to that component in tests, even if our test does not have an instance of the class that created the button. How and why we do this will be discussed in Chapter 8.

Summary

In this chapter, we introduced two classes, `UserStrings` and `ResourcesTester`, which together allow us to thoroughly test the resources used in our application. Other additional benefits are: very convenient methods for using resources, code completion for resource keys in our IDE, and the ability to weed out unused resources.

The sample application, 'Ikon Do It', uses this design pattern — see the source code for details.

6
Making Classes Testable with Interfaces

In the previous two chapters, we introduced specific programming and testing tools. In this chapter, we will be presenting a design pattern, namely, the appropriate use of Java interfaces. We will show how to apply this pattern to make small components or screens easy to test. We first look at how this applies in LabWizard, and then we will go into more detail by looking at an example that is given fully in the book source code.

The LabWizard Comment Editor

In the **LabWizard Knowledge Builder**, the main screen launches a **Comment Manager** which in turn launches a **Comment Editor**. Here is a view of the **LabWizard Knowledge Builder** application, showing a user in the process of editing one of the comments in a Knowledge Base.

Making Classes Testable with Interfaces

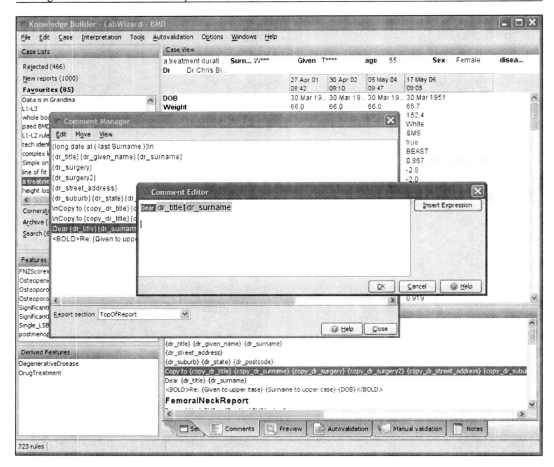

Obviously, there are lots of components working together here. Highest in the z-order is a **Comment Editor** dialog, which has been spawned from a **Comment Manager** dialog, itself arising from the **Knowledge Builder**. Now, the Comment Editor is a complex component and we need a thorough unit test for it. Obviously, there is some interaction between the Comment Manager and the Comment Editor, but we don't want to have to create instances of the former in our test of the latter. Likewise, we need to be able to test the Comment Manager without creating or even considering the Knowledge Builder from which it is launched.

The use of interfaces is the principal design pattern that enables us to test classes independently of the rest of the application. In this chapter, we illustrate this pattern using a much simpler example than the Comment Editor—namely the implementation of an on-screen 'wizard'.

The Wizard

A wizard is a GUI component that leads a user through the steps of some process. In our example, the main class is named Wizard and the steps are implemented by the class Step. The Step class provides a dialog with an informative title, an instructional label, a text field for the data entry, and a button for moving to the next step. Most real-life wizards have a **Previous** button too, but we'll leave that out.

At each step, the user enters some text and then presses the **Next** button. Step calls the Wizard's method nextPressed(String userInputForStep) to signal that the **Next** button has been pressed. This call-back allows Wizard to record the text that user has entered for that step.

Here is our first attempt at the Wizard class:

```java
public class Wizard {
    private int numberOfSteps;
    private JFrame frame = new JFrame();
    private int currentStepNumber = 1;
    private List<String> userEnteredText = new LinkedList<String>();
    public Wizard( int numberOfSteps ) {
        this.numberOfSteps = numberOfSteps;
        frame.setVisible( true );
    }
    public void showStep() {
        new Step( this, currentStepNumber ).show();
    }
    public void nextPressed( String userInputForStep ) {
        userEnteredText.add( userInputForStep );
        if (++currentStepNumber <= numberOfSteps) {
            showStep();
        } else {
            frame.dispose();
        }
    }
    public JFrame frame() {
        return frame;
    }
    public List<String> userEnteredText() {
        return userEnteredText;
    }
}
```

Now here is the code for `Step`. To make this example clearer, we have not used the `UserStrings` framework introduced in the last chapter.

```java
public class Step {
    private Wizard wizard;
    private int stepNumber;
    private JTextField textField = new JTextField( 20 );
    public Step( Wizard wizard, int stepNumber ) {
        this.wizard = wizard;
        this.stepNumber = stepNumber;
    }
    public void show() {
        String stepName = "Step " + stepNumber;
        final JDialog dialog =
                new JDialog( wizard.frame(), stepName, false );
        JButton nextButton = new JButton( "Next" );
        nextButton.setMnemonic( 'n' );
        nextButton.addActionListener( new ActionListener() {
            public void actionPerformed( ActionEvent e ) {
                wizard.nextPressed( textField.getText() );
                dialog.dispose();
            }
        } );
        JLabel label = new JLabel( "Enter info for " + stepName + ":" );
        Box box = Box.createHorizontalBox();
        box.add( label );
        box.add( textField );
        box.add( nextButton );
        dialog.getContentPane().add( box );
        dialog.pack();
        dialog.setName( stepName );
        dialog.setVisible( true );
    }
}
```

This is how the wizard appears:

This is clearly a very basic component, but it's sufficient for demonstrating the main concept.

A Test for Wizard

A simple test for the `Wizard` method `showStep()` is to show the first step, enter some text, activate the **Next** button, and repeat this for each `Wizard` step (there are five in this example). Once all the steps have been navigated, the test can check whether the `Wizard`'s version of the entered text matches what was typed in at each step. A complication is that we must both show the step and check any user-entered text in the event thread, for reasons explained in later chapters. This makes the test a lot more complex-looking than it really is. Later on, we'll look at easier ways of interacting with the Swing threads.

```java
public class WizardTest {
    private Wizard wizard;

    public boolean showStepTest() {
        //Create the wizard in a thread-safe manner,
        //and show the first step.
        final int numberOfSteps = 5;
        UI.runInEventThread( new Runnable() {
            public void run() {
                wizard = new Wizard( numberOfSteps );
                wizard.showStep();
            }
        } );

        //Type in some information at each step,
        //record it for later comparison,
        //and go to the next step.
        Cyborg cyborg = new Cyborg();
        List<String> expected = new LinkedList<String>();
        for (int i = 1; i <= numberOfSteps; i++) {
            String str = "Information for step " + i;
            expected.add( str );
            cyborg.type( str );
            cyborg.altChar( 'N' );
        }

        //Retrieve the entered text from the wizard
        //in a thread-safe manner.
        final List<List<String>> result = new
                                LinkedList<List<String>>();
        UI.runInEventThread( new Runnable() {
            public void run() {
                result.add( wizard.userEnteredText() );
            }
        } );
```

```
            //Compare the retrieved text with that expected
            Assert.equal( result.get(0), expected );
            return true;
        }
        //Other tests follow
    }
```

A Test for Step

A simple test for the `Step` method `show()` is to check the dialog's title, the text of the instructional label, and to check that once the **Next** button is pressed, the callback method to `Wizard` is in fact called with the text that has been entered at this step.

Here's some code that does this. Again, please overlook the hard-coded strings, and also the `EventThread` code needed for thread safety.

```
    public class StepTest {
        private Wizard wizard;
        private Step step;
        public boolean showTest() {
            //Create a wizard and from it a step.
            //Show the step.
            UI.runInEventThread( new Runnable() {
                public void run() {
                    wizard = new Wizard( 5 );
                    step = new Step( wizard, 1 );
                    step.show();
                }
            } );
            //Check the title and label of this step.
            final Dialog dialog = UI.findNamedDialog( "Step 1" );
            Assert.equal( UI.getTitle( dialog ), "Step 1" );
            assert UI.findLabelShowingText( "Enter info for Step 1:" )
                                                            != null;
            //Enter some text and press the Next button.
            String str = "Information for step 1";
            Cyborg cyborg = new Cyborg();
            cyborg.type( str );
            cyborg.altChar( 'N' );
            //Check that the entered text was received by the wizard.
            final List<List<String>> result = new LinkedList<List<String>
                                                                >();
            UI.runInEventThread( new Runnable() {
                public void run() {
                    result.add( wizard.userEnteredText() );
                }
            } );
```

```
            String receivedByWizard = result.get( 0 ).get( 0 );
            Assert.equal( receivedByWizard, str );
            UI.disposeOfAllFrames();
            return true;
        }
        //Other tests follow.
    }
```

The problem here is that the construction of a Step requires an instance of a Wizard, and hence the Wizard has become a necessary part of StepTest. In a production system, this could be a major headache as the Wizard class could be quite large with dependencies of its own, which would require still further classes to be part of the Step test. The unfortunate coupling between Wizard and Step has greatly increased the complexity of the StepTest and has made "bottom-up" testing impossible in this instance.

The answer is to decouple Wizard from Step by the use of an interface, which allows Step to be constructed without requiring a Wizard. This new version of Step is shown as StepV2:

```
    public class StepV2 {
        private int stepNumber;
        private JTextField textField = new JTextField( 20 );
        private Handler handler;
        public interface Handler {
            void nextPressed( String userInputForStep );
            JFrame frame();
        }
        public StepV2( Handler handler, int stepNumber ) {
            this.handler = handler;
            this.stepNumber = stepNumber;
        }
        public void show() {
            String stepName = "Step " + stepNumber;
            final JDialog dialog = new JDialog(
                    handler.frame(), stepName, false );
            JButton nextButton = new JButton( "Next" );
            nextButton.setMnemonic( 'n' );
            nextButton.addActionListener( new ActionListener() {
                public void actionPerformed( ActionEvent e ) {
                    handler.nextPressed( textField.getText() );
                    dialog.dispose();
                }
            } );
```

```
            JLabel label = new JLabel(
                    "Enter info for " + stepName + ":" );
            Box box = Box.createHorizontalBox();
            box.add( label );
            box.add( textField );
            box.add( nextButton );
            dialog.getContentPane().add( box );
            dialog.pack();
            dialog.setName( stepName );
            dialog.setVisible( true );
        }
    }
```

Note that we have bundled the methods that Step needed from Wizard into an interface StepV2.Handler, hereafter referred to simply as Handler. When we now test StepV2, rather than having to construct a Wizard, we construct a very simple implementation of Handler designed specifically for the test, as shown below:

```
public class StepV2Test {
    private JFrame frame;
    private String userInput;
    private HandlerForTest handler;
    private class HandlerForTest implements StepV2.Handler {
        public void nextPressed( String userInputForStep ) {
            userInput = userInputForStep;
        }
        public JFrame frame() {
            return frame;
        }
    }
    public boolean showTest() {
        //Create the frame and the handler.
        //Show the step.
        frame = UI.createAndShowFrame( "Test" );
        handler = new HandlerForTest();
        final StepV2 step = new StepV2( handler, 1 );
        UI.runInEventThread( new Runnable() {
            public void run() {
                step.show();
            }
        } );
        //Check the title and label of this step.
        final Dialog dialog = UI.findNamedDialog( "Step 1" );
        Assert.equal( UI.getTitle( dialog ), "Step 1" );
```

```
                assert UI.findLabelShowingText
                                ( "Enter info for Step 1:" ) != null;
                //Enter some text and press the Next button.
                String str = "Information for step 1";
                Cyborg cyborg = new Cyborg();
                cyborg.type( str );
                cyborg.altChar( 'N' );
                //Check that the entered text was received by the wizard.
                final String[] result = new String[1];
                UI.runInEventThread( new Runnable() {
                    public void run() {
                        result[0] = userInput;
                    }
                } );
                Assert.equal( str, result[0] );
                //Cleanup.
                UI.disposeOfAllFrames();
                return true;
            }
        //Other tests below.
    }
```

The new version of Wizard will of course have its own implementation of Handler, as shown in WizardV2:

```
    public class WizardV2 {
        private int numberOfSteps;
        private int currentStepNumber = 1;
        private List<String> userEnteredText = new LinkedList<String>();
        private JFrame frame = new JFrame();
        public WizardV2( int numberOfSteps ) {
            frame.setVisible( true );
            this.numberOfSteps = numberOfSteps;
        }
        public void showStep() {
            new StepV2( new StepHandler(), currentStepNumber ).show();
        }
        public List<String> userEnteredText() {
            return userEnteredText;
        }
        private class StepHandler implements StepV2.Handler {
            public void nextPressed( String userInputForStep ) {
                userEnteredText.add( userInputForStep );
                if (++currentStepNumber <= numberOfSteps) {
                    showStep();
                } else {
```

```
                frame.dispose();
            }
        }
        public JFrame frame() {
            return frame;
        }
    }
}
```

The design of the wizard has also been considerably improved by the use of the `Handler` interface. In the original `Wizard` class, the methods `nextPressed()` and `frame()` needed to be public so they could be accessed by `Step`. In `WizardV2` however, these methods are hidden within its private implementation `Handler`.

Handlers in LabWizard

Going back to the LabWizard example with which we started, we can see how this pattern improves the re-usability of our code. The Comment Editor can be created wherever a suitable handler can be supplied. This means that we can launch Comment Editors from various other components, rather than just from the Comment Manager, as would have been the case otherwise. This makes the software more task-oriented, rather than tool-oriented.

We were not using this design pattern in LabWizard from the outset, but now use it in all user interface components, as well as many server-side components, as it is the only way we have found of making them testable. In the earliest version of the Comment Editor, which by the way, was not unit-tested, the Comment Editor was created with an instance of the Knowledge Builder! When the time came to add new features to the Comment Editor, the only sensible way to unit test it was to break its coupling with the Knowledge Builder by using a handler interface.

Summary

The use of interfaces to decouple classes is a well-proven technique. It makes code simpler, more useful, and more testable. In particular, the pattern is of fundamental importance for testing user interface components, so we state it as a guideline:

[**Extreme Testing Guideline**: We should use interfaces to decouple our classes to improve both design and testability.]

7
Exercising UI Components in Tests

In this chapter, we introduce one of the key concepts in our user interface testing strategy—**UI wrapper classes**. The idea is that for each major user interface component, we create a class containing convenience methods for activating the controls in that component. This makes it extremely easy to write unit tests for the component. Further, because our wrappers do not actually need a reference to the component they control, they can be used in function tests. We'll see this use of UI wrappers in Chapter 17. Our introduction to UI wrappers will be via the unit tests for the **LabWizard** login screen.

The LabWizard Login Screen

Here is the login screen for **LabWizard**:

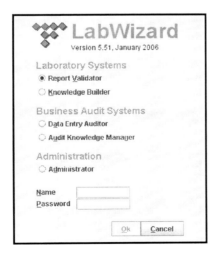

Every time a customer uses LabWizard, they begin with this component, so we want to make it as usable and bug-free as possible. Imagine how unimpressed a potential client would be if we started LabWizard on their machine and it looked like this:

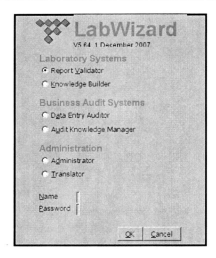

The figure shown above was the login screen for a pre-release version 5.64 of LabWizard, as it appeared on certain machines. We later wrote tests that replicated the incorrect sizing, and fixed the problem.

There are lots of tests we can write to ensure that our login screen is working properly. In terms of basic operation, we want to check that:

- When the **Ok** button is activated, the username, password, and application choice are passed on correctly.
- Any application choice can be made.
- After a successful login, the login screen vanishes.
- If the **Cancel** button is activated, the login screen vanishes and no data is passed on.

There are also a number of state properties to check:

- Initially, the **Report Validator** is selected, the username and password fields are blank, and the username field has focus.
- There is always exactly one radio button selected.
- The **Ok** button should not be enabled if either the name or the password field is empty.
- The password field masks entered passwords.
- All of the buttons and text fields are sensible sizes. This was a problem on certain machines when we added a new radio button in version 5.64.
- The LabWizard version number and release date are correct.

In terms of data verification, here are some things we want to ensure:

- If an invalid name or password is entered and **Ok** is pressed, a warning message is shown. When this message is acknowledged, the corresponding field should get the focus with all text selected.
- If the user selects an application for which they have insufficient privileges, a suitable warning message is shown.
- Exceptions thrown by the remote authentication server are handled smoothly.

Finally, we shouldn't neglect to test that the login screen satisfies certain usability criteria:

- The *Escape* key cancels the dialog.
- The *Enter* key is equivalent to pressing **Ok**.
- The mnemonics work.
- When a radio button is activated, the focus is passed to the name field so that the user can immediately start typing.
- Whenever the name or password field gets the focus, all of the text in that field should be selected.

To write tests such as these, we need programmatic tools for manipulating and reading the state of our user interfaces. These two things must be done in a thread-safe manner and they must be extremely easy to do, otherwise our tests will be hard to write and even harder to maintain. In this chapter, we look at the first of these requirements: the simulation of a user entering data with what we call 'UI Wrapper' classes.

The testing of details such as the mnemonics working may seem like a bit of an indulgence. However, if the testing is easy then it makes a lot of sense to do it. Why live with problems such as a mnemonic being used twice in a single screen, as in the following figure, if we can protect ourselves from them with just a few minutes work?

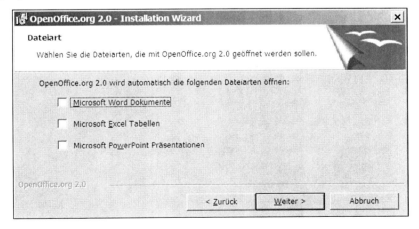

The Design of LoginScreen

Since we can't show much proprietary LabWizard code, we have re-implemented the login screen and its tests in the packages `jet.login` and `jet.login.test`.

Before looking at the way we test the login screen software, we should explain its structure, as this has some implications for unit testing. Our login screen is implemented by a class called—you guessed it—`LoginScreen`. The user interface is an undecorated `JFrame` that is displayed using a method called `show()`. The `show()` method blocks until either a successful login occurs or the login is cancelled.

When the user activates the **Ok** button, how will the `LoginScreen` pass the entered data to the remote server for verification? We could have a remote reference to the server as an instance variable for the `LoginScreen`, but that would mean that our unit tests for `LoginScreen` would be very complicated. Instead, we make use of the design pattern introduced in the previous chapter and represent this remote server by an interface that can be implemented easily in test classes.

A few helper classes for `LoginScreen` are:

Class	Description
`LoginInfo`	Encapsulates a username, password, and application choice.
`ApplicationChoice`	An enum with a value for each of the available applications.
`LoginResult`	An inner enum of `LoginScreen`. The values represent possible results of a logon attempt: success, failure because the user is unknown, failure because the password is wrong, and failure because the user has insufficient privileges for the requested application.
`Handler`	Inner interface of `LoginScreen` that represents the authentication server.

In terms of error handling, the code presented here is simple and just prints stack traces to the console. The real implementation logs them.

Here then is an outline of the code for `LoginScreen`:

```
public class LoginScreen {
    public static enum LoginResult {
        ACCEPTED, UNKNOWN_USER, WRONG_PASSWORD,
        INSUFFICIENT_PRIVILEGES
    }
    public interface Handler {
        LoginResult login( LoginInfo info );
    }
```

```
private Handler handler;
private JFrame frame;
//The current choice of application, set from the radio buttons.
private ApplicationChoice chosen;
//Other components.
.....
public LoginScreen( Hander handler ) {
    this.handler = handler;
}
/**
 * Shows the login screen, in a swing thread-safe manner,
 * then waits for a successful login or cancellation.
 */
public ApplicationChoice show() {
    //Create and show the screen, in the Swing event thread.
    try {
        SwingUtilities.invokeAndWait( new Runnable() {
            public void run() {
                buildUI();
                frame.setVisible( true );
                frame.toFront();
                userNameField.requestFocusInWindow();
            }
        } );
    } catch (Exception e) {
        e.printStackTrace();
    }
    //Wait for the login or cancellation.
    synchronized (this) {
        try {
            wait();
        } catch (InterruptedException ie) {
            ie.printStackTrace();
        }
    }
    //Return the selection, or null if cancelled.
    return chosen;
}
//Called after a successful login or when cancelled.
private void disposeFrameAndWakeLaunchingThread() {
    frame.dispose();
    synchronized (this) {
        notify();
    }
}
//Called when ok pressed.
```

```
        private void ok() {
            String name = userNameField.getText();
            String password = passwordField.getText();
            LoginInfo loginInfo =
                    new LoginInfo( name, password, chosen );
            LoginResult result = handler.login( loginInfo );
            switch (result) {
                case ACCEPTED:
                    disposeFrameAndWakeLaunchingThread();
                    break;
                case UNKNOWN_USER:
                    //For negative results, show an appropriate message.
                    ....
            }
        }
        //Lots of stuff omitted. In particular, the radio buttons set
        //the 'chosen' field if selected. The 'cancel' action nulls it.
        ....
}
```

Some readers may be surprised that LoginScreen has a JFrame, rather than extending that class. We're following the **Liskov Substitution Principle**, which states that a class Y should extend a class X only if whenever we can use an X, we can use a Y. Since LoginScreen cannot generally take the place of JFrame, LoginScreen does not extend JFrame. We have found that extending Swing classes is normally not a good idea, apart from a few clear exceptions like renderers. Delegation to Swing components is a much safer and cleaner design pattern. By the way, the substitution principle explains the seemingly unintuitive rules, mentioned in Chapter 1. These rules state that the pre-conditions to a method re-definition must be OR-ed with those of its ancestor methods, and that post-conditions must be AND-ed as we go down an inheritance chain. See *"Object-Oriented Software Construction"*, Second Edition, by Bertrand Meyer (Prentice Hall, 1997) for details.

UI Wrappers

To implement the tests listed at the start of this chapter, we will need easy ways of selecting different radio buttons, entering names and passwords, and activating the **Ok** and **Cancel** buttons.

We could do this with methods in LoginScreenTest itself. However, we prefer to encapsulate this functionality in what we call a UI Wrapper class that can then be used easily by other tests, particularly function tests.

The wrapper for the login screen will be called `UILoginScreen` and will declare the following public methods:

```
public class UILoginScreen {
    public void chooseApplication( ApplicationChoice choice ) {...}
    public void enterName( String name ) {...}
    public void enterPassword( String password ) {...}
    public void ok() {...}
    public void cancel() {...}
}
```

How can we implement these methods? One way would be to provide accessor methods to the various user interface components from within `LoginScreen` itself, and create `UILoginScreen` instances from `LoginScreen` instances:

```
//Possible (but flawed) implementation of UILoginScreen.
public class UILoginScreen {
    private LoginScreen loginScreen;
    public UILoginScreen( LoginScreen loginScreen ) {
        this.loginScreen = loginScreen;
    }
    //Methods for entering data.
    public void chooseApplication( ApplicationChoice choice ) {
        buttonForChoice( choice ).setSelected( true );
    }
    //...Other methods for entering data.
}
```

There are several problems with this approach.

- Firstly, we will be creating a whole lot of new methods in our production code just for testing. Sometimes there is no alternative to writing such methods, but we would rather not do it.
- Secondly, we will always need a `LoginScreen` with which to create our UI Wrapper. This will be fine in the unit tests, but later on, we want to create more complex function tests where we will not have access to a `LoginScreen` object.
- Finally, the methods shown will be setting and reading Swing objects from our main test thread, which is not the Swing event thread. This will lead to 'flaky' tests caused by concurrency issues. In other words, this design is not thread-safe.

Our actual implementation of `UILoginScreen` will be based on the principle that all data entry is to be performed by the generation of events that mimic user actions, such as typing and mouse gestures. Each component should have a mnemonic that either activates it (in the case of `JButton`) or gives it the focus (`JTextField`, `JList`, and so on). This is so entering data programmatically is no problem, especially with `Cyborg` on our side.

 Extreme Testing Guideline: In user interface tests, the state of UI components should only be changed by using mouse and keyboard events, not by calling methods on the UI component directly.

The exception to this rule is that our tests will generally have to actually create and display the components (technically a state change) before they can be exercised via mouse and keyboard events.

The Correct Implementation of UILoginScreen

Recall from Chapter 5 that we will typically create buttons by using a `UserStrings` instance and a constant:

```
public class LoginScreen {
    private UserStrings us = LoginUserStrings.instance();
    ....
    okButton = us.createButton( okAction, LoginUserStrings.OK );
    ...
}
```

This will give the button a mnemonic

```
us.mnemonic( LoginUserStrings.OK );
```

from the appropriate properties file.

So if our `UILoginScreen` class has a `UserStrings` instance and a `Cyborg`, it can easily activate the buttons using the mnemonic:

```
public class UILoginScreen {
    private LoginUserStrings us = LoginUserStrings.instance();
    private Cyborg cyborg = new Cyborg();
    public void ok() {
        cyborg.altChar( us.mnemonic( LoginUserStrings.OK ) );
    }
    ...
}
```

We can also use this approach for entering text. First, we set the focus to the field in which the text is supposed to be entered, which is done by activating the mnemonic for the label associated with the field. The field should be set up so that selecting it selects all text, so we can simply type in our intended value. If we mean to enter a blank value, we just delete the selected text:

```
public class UILoginScreen {
    ...
    public void enterName( String name ) {
        enterNameOrPassword( name, LoginUserStrings.USER_NAME );
    }...
    private void enterNameOrPassword( String str, String key ) {
        //Get focus to the appropriate field.
        cyborg.altChar( us.mnemonic( key ) );
        //All should now be selected, so just type in the data.
        if (str.equals( "" )) {
            cyborg.delete();//Just delete what is already there.
        } else {
            cyborg.type( str );
        }
    }
    ...
}
```

The other data entry methods are implemented in a similar fashion, and there is no need for `UILoginScreen` to have a reference to an actual `LoginScreen`.

A Handler Implementation for Unit Testing

To construct a `LoginScreen`, we need a `LoginScreen.Handler` instance. In our unit test, we will use an implementation that is designed for exercising the `LoginScreen`. We want our test implementation of the handler to be able to imitate any behavior expected of the real implementation. This includes:

- Accepting a login.
- Rejecting a login because of:
    - Unknown user name
    - Incorrect password
    - Insufficient user privileges for the requested application.
- Throwing an exception (which might be a wrapper for a `RemoteException` if, for example, the authentication server is not available).

We also need our handler to record the data passed to it and to be able to check this against expected values. Here then is the test handler implementation, as an inner class of `LoginScreenTest`:

```
public class LoginScreenTest {
    ...
    //Test implementation of the Handler.
    private class TestHandler implements LoginScreen.Handler {
        //The data most recently passed in by the LoginScreen.
        LoginInfo enteredData;
        //The value for login() to return.
        LoginScreen.LoginResult result;
        //The exception to be thrown by login()
        RuntimeException toThrow;
        public LoginScreen.LoginResult login( LoginInfo info ) {
            enteredData = info;
            if (toThrow != null) throw toThrow;
            return result;
        }
        void checkData( String name, String password,
                    ApplicationChoice choice ) {
            assert enteredData.userName().equals( name ) :
                "Expected: " + name + ", got: "
                    + enteredData.userName();
            assert enteredData.password().equals( password ):
                "Expected: " + password + ", got: "
                    + enteredData.password();
            assert enteredData.chosenApplication().equals( choice )
                : "Expected: " + choice + ", got: "
                    + enteredData.chosenApplication();
        }
    }
}
```

Setting Up our Tests

The unit test will have short, very easy to maintain methods, each of which tests just one thing and can be run on its own. A tried and true way of doing this is to have each method of the form:

```
public boolean aTest() {
    init();
    //Test body.....
    cleanup();
    return true;
}
```

The `init()` method creates a handler instance, uses it to build and show a `LoginScreen`, and creates a `UILoginScreen`. The `cleanup()` method destroys any frames or other resources created in the tests. Our preference is to explicitly call methods for both initializing and cleaning up after the tests. This is because we often need to customize these processes. There might be several such methods in a test, or they might take parameters. By choosing between our initialization methods, we can mimic particular situations. We have found the JUnit approach, of having `setUp()` and `tearDown()` called automatically, too inflexible.

The body of the test will throw an assertion failure if any problems are encountered. If no problems are found, the `LoginScreen` does not have the possible defect being tested for, and `true` is returned. Since assertion errors are thrown to indicate test failures, the return value is redundant. However, it is required by the test framework—see the discussions in Chapters 2 and 4.

One complication we have is that the `show()` method of `LoginScreen` blocks, so it cannot be called by our main test thread. Instead, we will spawn a new thread that shows the login screen. We could make this thread an anonymous inner class. However, we want our tests to check that the thread that shows a login screen is woken up when a login or cancel occurs. For this reason, we will create a special launcher class:

```
public class LoginScreenTest {
    ...
    private class Launcher extends Thread {
        public boolean isAwakened;
        public Launcher() {
            super( "Launcher" );
            setDaemon( true );
        }
        public void run() {
            ls.show();//This is Swing thread-safe.
            isAwakened = true;
        }
    }
}
```

Here are the `init()` and `cleanup()` methods:

```
public class LoginScreenTest {
    private TestHandler handler;
    private LoginScreen ls;
    private UILoginScreen uils;
    private Launcher launcher;
    private LoginUserStrings us = LoginUserStrings.instance();
```

```
    ...
    private void init() {
        handler = new TestHandler();
        handler.result = LoginScreen.LoginResult.ACCEPTED;
        ls = new LoginScreen( handler );
        launcher = new Launcher();
        launcher.start();
        //Wait for the login screen to show.
        ...
        //Now that the LoginScreen is showing,
        //we can create its UI wrapper.
        uils = new UILoginScreen();
    }
    private void cleanup() {
        //Close the login screen if it's still up.
        ...
    }
    ...
}
```

Our First Test

After all this preparation (about 80 lines of code and comments), we are ready to write the actual tests. Let's look at a test that checks the behavior of LoginScreen when a user selects an application for which they do not have sufficient privileges. What we want to check is that a helpful message is displayed to the user, and that when this message is acknowledged, the login screen is showing the name and password first entered.

```
public boolean invalidAdministratorLoginTest() {
    init();
    //Set up the handler to fail because the user does
    //not have privileges for the chosen application.
    handler.result
        = LoginScreen.LoginResult.INSUFFICIENT_PRIVILEGES;
    uils.enterName( "Harry" );
    uils.enterPassword( "Firebolt" );
    uils.chooseApplication( ApplicationChoice.ADMINISTRATOR );
    uils.ok();

    //Sanity check that the data has been passed in.
    handler.checkData( "Harry", "Firebolt",
                       ApplicationChoice.ADMINISTRATOR );
```

```
    //Check that the expected error message is showing.
    String expectedError = us.message(
        LoginUserStrings.INSUFFICIENT_PRIVILEGES_MSG2,
        "Harry",
         us.label( ApplicationChoice.ADMINISTRATOR.toString() ) );
    ...
    //Acknowledge the error message.
    uils.cyborg().activateFocussedButton();

    //Check that the error message is no longer showing.
    ...
    //Check that the Login Screen is still showing, with the
    //same data entered. Do this by activating ok and checking
    //the passed in data.
    handler.enteredData = null;//Clear out previous values first.
    uils.ok();
    handler.checkData( "Harry", "Firebolt",
                            ApplicationChoice.ADMINISTRATOR );
    //Check that the correct error message is given again.
    ...
    cleanup();
    return true;
}
```

There are a few things to note here.

- First of all, we have left out the code for checking the error message. We'll see how to do this in Chapter 8.
- Secondly, in this test, we wanted to check that after the error message had been acknowledged, the name and password fields were still showing the values they had prior to **Ok** being pressed. We did this check in a roundabout way—by activating **Ok** again and examining the values sent to the handler. An alternative approach would be to check the values in the JTextField components directly. We could do this, but it would require techniques developed later in Chapters 8 and 10.
- Finally, **this test is not thread-safe**. The reason is that we are querying the handler in our test thread for values set in the Swing event thread. We will look at how this can be fixed in Chapter 9.

To cut down on the boilerplate code, we can add a 'higher order' method to
`UILoginScreen`:

```
public void login( String name, String password,
                   ApplicationChoice choice ) {
    enterName( name );
    enterPassword( password );
    chooseApplication( choice );
    ok();
}
```

With this refactoring, our test becomes pretty lean in that it largely consists of single lines that directly express the logic of the test. That is, it is pretty clear how to go from the specification of the test to the code for it.

To see the fully refactored test, look in the source code. The test is called `invalidAdministratorLogin2Test()`.

Should we add tests for insufficient privileges for other application choices? What we are really testing is that the error message given is tailored to the `ApplicationChoice`. What are the consequences of a bug in this code? Should we spend the time writing different tests? In general, these kinds of questions need to be considered in the context of the project we are working on. Every test we write is time not spent testing something else, or adding new functionality to our product. In this case, however, we should generalize the test as it is so easy to do so:

```
public boolean insufficientPrivilegesTest() {
    for (ApplicationChoice ac : ApplicationChoice.values()) {
        checkLoginWithInsufficientPrivilegesForApp( ac );
    }
    return true;
}
private void checkLoginWithInsufficientPrivilegesForApp(
        ApplicationChoice ac ) {
    //Refactored from invalidAdministratorLoginTest.
    ...
}
```

Further Tests

We have not yet tested all of the things we set out to. There is no test that the **Ok** button is enabled only if a user name and password are entered. There is no test that the thread that launched the login screen is woken up when a cancellation occurs. These tests, and others, require techniques that are developed in the next chapter. The tests actually have been implemented and can be browsed with the source code.

Some Implicit Tests

At the start of this chapter, we boasted that we'd make it totally trivial to check that all the mnemonics in the login screen are well-defined. Our wrapper class UILoginScreen does this implicitly by using mnemonics to activate or select the various controls. So if we've got a test in which we use the wrapper to activate the **Ok** button, and if the test passes, we know that the **Ok** button was indeed activated using its mnemonic.

Similarly, the wrapper's implementation of enterName() and enterPassword() would only work if the text fields selected all text upon receiving the focus.

If we want more explicit and traceable tests of usability, for example, if we have to show that certain guidelines are being followed, we can always write them.

Other User Interfaces

The methods in UILoginScreen for choosing an application, entering the name and password, and so on, only rely on the fact that the LoginScreen component responds to keyboard shortcuts. If we had another implementation of LoginScreen, but one that still respected these keyboard bindings, we could use the same UI wrapper class for testing it. For instance, we might create a JavaServer Faces login screen that looked and behaved much like LoginScreen, and then manipulate it in tests using UILoginScreen.

Summary

In this chapter, we've seen how to write UI wrapper classes that make the programmatic manipulation of user interfaces straightforward, thread-safe, and free from boilerplate code.

8
Showing, Finding, and Reading Swing Components

In the previous chapter, we implemented our first test of a Swing user interface, the LabWizard login screen. However, we glossed over several important points. For example, we did not provide a test that the **Ok** button of a login screen was enabled if and only if a user name and password had been entered. Also, we did not look at how we could check that a particular error or warning message was being shown.

In order to write such tests, we need a way of finding individual components, such as a particular button, or some label showing specific text. We also need techniques for reading the state of components, such as whether they are enabled, or what text they are showing.

The underlying difficulty here is that our test thread is not the Swing event thread, yet needs to interact with Swing components. The Swing library is not thread-safe, so any manipulation or interrogation of user interface components must be done from within the Swing event thread. In the previous chapter, we used UI Wrapper classes to manipulate Swing components with keyboard events. This technique is thread-safe because the `Cyborg` instance that generates the events waits for each event to be processed before generating the next.

Another issue we have not dealt with properly is how to actually show a component so that we can test it. For the login screen, this was pretty easy because that component was designed to be invoked from the JVM main thread, rather than the Swing event thread. In general, though, our Swing components are not built like this, we need a way of showing them safely.

The exact manner in which a component can be displayed for testing is specific to that component. However, a lot of user interfaces need a `JFrame`, either as a component container or dialog owner, so in this chapter we will look at how to safely show a `JFrame`. We will also develop methods to find components, and to safely read their state. In fact, we will be building a library of helper methods, which are static methods in a class called `UI`.

The techniques covered in this chapter are very important in testing user interfaces. If we do not pay proper attention to concurrency issues, our tests will be 'flaky'—they will almost always run correctly, but a small percentage of runs will fail. A small chance of failure for a single test means certain failure for a batch of hundreds or thousands of tests. We require our tests to run automatically as part of our build process, so we have to get the threading issues right.

Setting Up User Interface Components in a Thread-Safe Manner

A safe way of programmatically manipulating Swing components from a thread other than the Swing event thread is via the methods `invokeLater (Runnable r)` and `invokeAndWait(Runnable r)`. These are both declared in the `javax.swing.SwingUtilities` class.

This mucking around with `Runnable`s is very tiresome. To ease the pain, we've developed a library of handy methods for setting up and reading the state of Swing components in a thread-safe manner. As we will forever be setting up frames, here is a method for doing it safely:

```
public class UI {
    /**
     * Use the event thread to show a frame. When this method has
     * returned the frame will be showing and to the front.
     */
    public static void showFrame( final JFrame frame ) {
        runInEventThread( new Runnable(){
            public void run() {
                frame.setVisible( true );
                frame.toFront();
            }
        });
    }
    //...
}
```

This in turn makes use of a wrapper method for `invokeAndWait()`:

```
/**
 * Handy wrapper for <code>SwingUtilities.invokeAndWait()</code>
 * that does not try to use the event thread if that is the
 * calling thread.
 */
```

```
public static void runInEventThread( final Runnable r ) {
    if (SwingUtilities.isEventDispatchThread()) {
        r.run();
    } else {
        try {
            SwingUtilities.invokeAndWait( r );
            new Cyborg().robot().waitForIdle();
        } catch (Exception e) {
            e.printStackTrace();
            assert false : e.getMessage();
        }
    }
}
```

Our method for showing a frame can be used within the initialization methods of any tests. For example, in a test of a particular user interface component we will need to create the component, put it in a frame, and show the frame:

```
//Code to create the canvas sub-class
canvas = ...;
//A frame to show it in.
JFrame frame = ...;
//Put the canvas in the frame and show it.
Runnable creater = new Runnable() {
    public void run() {
        frame = new JFrame( "IkonCanvasTest" );
        frame.add( canvas.component() );
        frame.pack();
    }
};
UI.runInEventThread( creater );
UI.showFrame( frame );
```

We can be certain after running this code that our component is showing in a frame and has focus. It is now ready to receive mouse and keyboard events, or to have its state investigated. (This code is from the unit test for a component called an IkonCanvas, which is part of the 'Ikon Do It' application. See the package jet.ikonmaker in the source code for more details.)

This careful setting up of the component in a frame avoids occasional test failures caused by mouse and keyboard events not being received due to the component not yet having focus.

Finding a Component

The `LoginScreen` class has no methods that give access to the buttons, labels, and text fields in the login screen that it defines. How then can we get access to these objects in order to test the usability of our login screen?

There is a method called `getComponents()` which gives access to all components in a `java.awt.Container`. We can use this recursively to search for a component that is located somewhere in a `java.awt.Window` or dialog. It is also possible to get a list of all windows owned by some particular root `Frame`. Finally, there is a method `Frame.allFrames()` that gives access to all frames in a JVM. Together, these methods will allow us to find any component, as long as we have a means of identifying it.

Our class `UI` implements this search algorithm to provide methods for finding components. An inner interface, `SearchCriterion`, encapsulates the means of recognizing the component we want:

```java
public final class UI {
    public static interface ComponentSearchCriterion {
        public boolean isSatisfied( Component component );
    }
    ...
}
```

The method for searching amongst a group of components for a particular one is then:

```java
    private static Component findComponentAmongst(
            final Component[] components,
            final ComponentSearchCriterion isMatch ) {
        for (int i = 0; i < components.length; i++) {
            final Component component = components[i];
            if (isMatch.isSatisfied( component ) ) {
                return component;
            }
            if (component instanceof Container) {
                final Component recurse =
                    findComponentAmongst( ((Container) component).
                                        getComponents(), isMatch );
                if (recurse != null) {
                    return recurse;
                }
            }
        }
        return null;
    }
```

We can use this in the following method to search within a window for a component, no matter how deeply nested the component is:

```
public static Component findComponentInWindow(
            Window w, ComponentSearchCriterion criterion ) {
    return findComponentAmongst( w.getComponents(), criterion );
}
```

The most general method for finding a component is to search through all of the frames in the JVM, and all sub-windows of the frames:

```
private static Component findComponentInSomeFrame(
        final ComponentSearchCriterion criterion ) {
    final Component[] resultHolder = new Component[1];
    runInEventThread( new Runnable() {
        public void run() {
            Frame[] allFrames = Frame.getFrames();
            for (Frame frame : allFrames) {
                if (!frame.isShowing()) {
                    continue;
                }
                Component result = findComponentInWindow(
                        frame, criterion );
                if (result != null) {
                    resultHolder[0] = result;
                    return;
                } else {
                    Window[] subWindows = frame.getOwnedWindows();
                    for (Window subWindow : subWindows) {
                        result = findComponentInWindow(
                                subWindow, criterion );
                        if (result != null) {
                            resultHolder[0] = result;
                            return;
                        }
                    }
                }
            }
        }
    } );
    return resultHolder[0];
}
```

As an example of how to use this method, consider this code for finding a label, based on the text it is showing:

```
public static JLabel findLabelShowingText( final String text ) {
    return (JLabel) findComponentInSomeFrame(
            new ComponentSearchCriterion() {
        public boolean isSatisfied( Component comp ) {
            if (comp instanceof JLabel) {
                return text.equals( ((JLabel) comp ).getText() );
            }
            return false;
        }
    } );
}
```

We'll give more examples of this technique later. But for now, let's clear up one of the details omitted from the unit test developed in the previous chapter.

Testing Whether a Message is Showing

In our unit test for `LoginScreen` presented in the last chapter, we skipped the details of how we could check that a particular error message was showing.

A `JOptionPane` shows messages using `JLabel`. Each line of the message is given its own label, and these are laid out in a column.

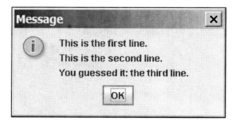

Therefore, assuming that the error message will be presented in a `JOptionPane`, we can use the `UI.findLabelShowingText()` method to check whether the error message is showing. Here are the missing lines from `LoginScreenTest`:

```
    ...
    //Check that the expected error message is showing.
    String expectedError = us.message(
            LoginUserStrings.INSUFFICIENT_PRIVILEGES_MSG2,
            "Harry",
            us.label( ApplicationChoice.ADMINISTRATOR.toString() ) );
```

```
        assert UI.findLabelShowingText( expectedError ) != null;
        //Acknowledge the error message.
        uils.cyborg().activateFocussedButton();
        //Check that the error message is no longer showing.
        assert UI.findLabelShowingText( expectedError ) == null;
        ...
```

Searching for Components by Name

Searching for a `JLabel` that has the text of an expected error message is a good way of checking that a `JOptionPane` is showing that error message. The success of the search is equivalent to whether or not the message is showing. What can we do when the component we are looking for does not display any text? What can we do when the component displays frequently used text that might be in several components, such as **Ok** or **Cancel**?

One way to handle this is to name the components we need to search for. Our `UI` class provides a method to search for any component by name:

```
    /**
     * The component having the given name, or null if none is found.
     */
    public static Component findNamedComponent( final String name ) {
        return findComponentInSomeFrame(
                new ComponentSearchCriterion() {
                    public boolean isSatisfied( Component component ) {
                        return name.equals( component.getName() );
                    }
                } );
    }
```

In the previous chapter we showed the manipulative methods of `UILoginScreen`, but there are also methods that give access to the sub-components:

```
    public class UILoginScreen {
        ... See previous chapter for manipulative methods.
        public JTextField nameField() {
            return (JTextField) UI.findNamedComponent(
                        LoginUserStrings.USER_NAME );
        }
        public JPasswordField passwordField() {
            return (JPasswordField) UI.findNamedComponent(
                        LoginUserStrings.PASSWORD );
        }
        public JButton okButton() {
```

```
            return (JButton) UI.findNamedComponent(
                            LoginUserStrings.OK );
    }
    public JButton cancelButton() {
        return (JButton) UI.findNamedComponent(
                        LoginUserStrings.CANCEL );
    }
    public JRadioButton buttonForChoice(
                        ApplicationChoice choice ) {
        return (JRadioButton) UI.findNamedComponent(
                            choice.toString() );
    }
    public JFrame frame() {
        return frame;
    }
}
```

In `LoginScreen`, we named the text fields with the `LoginUserStrings` constants for their associated labels. An alternative would have been to put a public constant for each of them into the `LoginScreen` class itself. We generally create buttons with a `UserStrings` (see Chapter 5) factory method: `public JButton createJButton ( Action a, String key )`. This method names the button it creates with the given key.

In general, naming the components we want to interrogate is so useful that we recommend that it always be done:

[**Extreme Testing Guideline**: Any component that might need to be found in a test should be given a name that is a public constant in some associated class.]

Reading the State of a Component

Once we have found a component, the correct way to read its state is from within the Swing event thread. Our `UI` class has many methods for doing this. For example, to see whether or not a component is enabled:

```
    /**
     * Safely read the enabled state of the given component.
     */
    public static boolean isEnabled( final Component component ) {
        final boolean[] resultHolder = new boolean[]{false};
        runInEventThread( new Runnable() {
            public void run() {
                resultHolder[0] = component.isEnabled();
```

```
            }
        } );
        return resultHolder[0];
    }
```

Using this method, we can write the long-promised test for the state of the login dialog **Ok** button:

```
    public boolean okButtonEnabledStateTest() {
        init();
        assert!UI.isEnabled( uils.okButton() );
        uils.enterName( "a" );
        assert!UI.isEnabled( uils.okButton() );
        uils.enterPassword( "a" );
        assert UI.isEnabled( uils.okButton() );
        uils.enterPassword( "" );
        assert!UI.isEnabled( uils.okButton() );
        uils.enterPassword( "a" );
        assert UI.isEnabled( uils.okButton() );
        uils.enterName( "" );
        assert!UI.isEnabled( uils.okButton() );
        uils.enterName( "a" );
        assert UI.isEnabled( uils.okButton() );

        cleanup();
        return true;
    }
```

Case Study: Testing Whether an Action Can Be Cancelled

Let's suppose we are testing a function that exports an image as a file. We want our users to be warned when the file they have chosen already exists. How do we write a test to check that the warning message is showing? How do we test that if they cancel, the action is aborted, and that if they choose to go ahead with it, the file is over-written as intended?

The following example from the unit tests for 'Ikon Do It' shows how to do this. In the code that follows, `ui` is a UI Wrapper (see Chapter 7) for the main 'Ikon Do It' application frame, which is defined in the class `IkonMaker`. The `exportWarning` `String` is the warning message that should be on display

```
    public boolean exportTest() throws IOException {
        //Set up the IkonMaker main frame, delete and files in
        //the export directory, and so on.
```

```
        init();
        ui.createNewIkon( "test", 16, 16 );
        File dest = new File( Files.tempDir(), "FirstExport.png" );
        assert !dest.exists();
        ui.export( dest );
        assert dest.exists();
        assert dest.isFile();
        //Check the saved image....
        ...
        //Now try to export it again.
        //First we will cancel. We'll check that the
        //last modified time of the file has not changed.
        long lastModified = dest.lastModified();
        //Pause because the timing resolution is
        //pretty rough on some platforms.
        Waiting.pause( 100 );//See Chapter 12.
        ui.export( dest );
        //Check that a dialog with the expected
        //warning message is showing.
        assert UI.findLabelShowingText( exportWarning ) != null;
        //Focus is on yes, tab to no.
        cyborg.tab();
        cyborg.space();
        //We will check that the file has not been over-written
        //by looking at the last time it was modified.
        assert dest.lastModified() == lastModified;

        //This time, overwrite.
        ui.export( dest );
        assert UI.findLabelShowingText( exportWarning ) != null;
        //Focus is on yes.
        cyborg.space();
        assert dest.lastModified() > lastModified;
        //Check that the saved image is still ok.
        ....
    }
```

This is typical of the way we can test the behavior of our user interfaces. More examples can be found in the source code.

Chapter 8

The Official Word on Swing Threading

For more information on the thread safety issues we've covered here, look to: `http://java.sun.com/products/jfc/tsc/articles/threads/threads1.html`.

Here is a small extract from this document:

> **A few methods are thread-safe:** *In the Swing API documentation, thread-safe methods are marked with this text:*
>
> **This method is thread safe, although most Swing methods are not**.
>
> *An application's GUI can often be constructed and shown in the main thread. The following typical code is safe, as long as no components (Swing or otherwise) have been realized:*
>
> ```
> public class MyApplication {
> public static void main(String[] args) {
>     JFrame f = new JFrame("Labels");
>     // Add components to
>     // the frame here...
>     f.pack();
>     f.show();
>     // Don't do any more GUI work here...
>     }
> }
> ```
>
> *All the code shown above runs on the* main *thread. The* f.pack() *call realizes the components under the* JFrame. *This means that, technically, the* f.show() *call is unsafe and should be executed in the event-dispatching thread. However, as long as the program doesn't already have a visible GUI, it's exceedingly unlikely that the* JFrame *or its contents will receive a* paint() *call before* f.show() *returns. Because there's no GUI code after the* f.show() *call, all GUI work moves from the main thread to the event-dispatching thread, and the preceding code is, in practice, thread-safe.*

If we were running a single test, we might be prepared to live with the low risk of adverse behavior caused by an unsafe call to `show()`. However, in the testing of a complex system, we might already have a visible GUI, we almost certainly have just disposed of one, and we aim to run hundreds or even thousands of tests. Given these circumstances, and our experience of many hours fixing "flaky" tests, we recommend total adherence to thread-safe programming, and the UI class is there to make it as painless as possible.

Summary

In this chapter, we have seen how to safely launch components at the start of a test, and how to read their state during the test itself. We've also seen how to get hold of individual components by searching for them amongst all components in the JVM. To make this kind of component search easy, we need to name all of our components.

9

Case Study: Testing a 'Save as' Dialog

In Chapter 7 we presented the idea of UI Wrappers for components as a way of safely and easily manipulating them in tests. In Chapter 8 we developed this further with specific techniques for reading the state of our user interfaces. We will now draw these two strands together by studying in detail the test for an extremely simple user interface component. Although the component that we'll be testing in this chapter is simple, it will still allow us to introduce a few specific techniques for testing Swing user interfaces. It will also provide us with an excellent opportunity for showing the basic infrastructure that needs to be built into these kinds of tests.

In this chapter, we are taking a break from LabWizard and will use an example from our demonstration application, Ikon Do It. The code from this chapter is in the packages `jet.ikonmaker` and `jet.ikonmaker.test`.

The Ikon Do It 'Save as' Dialog

The 'Ikon Do It' application has a **Save as** function that allows the icon on which we are currently working to be saved with another name. Activating the **Save as** button displays a very simple dialog for entering the new name. The following figure shows the 'Ikon Do It' **Save as** dialog.

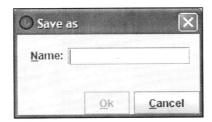

Case Study: Testing a 'Save as' Dialog

Not all values are allowed as possible new names. Certain characters (such as '*') are prohibited, as are names that are already used.

In order to make testing easy, we implemented the dialog as a public class called `SaveAsDialog`, rather than as an inner class of the main user interface component. We might normally balk at giving such a trivial component its own class, but it is easier to test when written this way and it makes a good example. Also, once a simple version of this dialog is working and tested, it is possible to think of enhancements that would definitely make it too complex to be an inner class. For example, there could be a small status area that explains why a name is not allowed (the current implementation just disables the **Ok** button when an illegal name is entered, which is not very user-friendly).

The API for `SaveAsDialog` is as follows. Names of icons are represented by `IkonName` instances. A `SaveAsDialog` is created with a list of existing `IkonNames`. It is shown with a `show()` method that blocks until either **Ok** or **Cancel** is activated. If **Ok** is pressed, the value entered can be retrieved using the `name()` method. Here then are the public methods:

```java
public class SaveAsDialog {
    public SaveAsDialog( JFrame owningFrame,
                         SortedSet<IkonName> existingNames ) { ... }
    /**
     * Show the dialog, blocking until ok or cancel is activated.
     */
    public void show() { ... }
    /**
     * The most recently entered name.
     */
    public IkonName name() { ... }
    /**
     * Returns true if the dialog was cancelled.
     */
    public boolean wasCancelled() { ... }
}
```

Note that `SaveAsDialog` does not extend `JDialog` or `JFrame`, but will use delegation like `LoginScreen` in Chapter 7. Also note that the constructor of `SaveAsDialog` does not have parameters that would couple it to the rest of the system. This means a handler interface as described in Chapter 6 is not required in order to make this simple class testable.

The main class uses `SaveAsDialog` as follows:

```
private void saveAs() {
    SaveAsDialog sad = new SaveAsDialog( frame,
            store.storedIkonNames() );
    sad.show();
    if (!sad.wasCancelled()) {
        //Create a copy with the new name.
        IkonName newName = sad.name();
        Ikon saveAsIkon = ikon.copy( newName );
        //Save and then load the new ikon.
        store.saveNewIkon( saveAsIkon );
        loadIkon( newName );
    }
}
```

Outline of the Unit Test

The things we want to test are:

- Initial settings:
    - The text field is empty.
    - The text field is a sensible size.
    - The **Ok** button is disabled.
    - The **Cancel** button is enabled.
    - The dialog is a sensible size.
- Usability:
    - The *Escape* key cancels the dialog.
    - The *Enter* key activates the **Ok** button.
    - The mnemonics for **Ok** and **Cancel** work.
- Correctness. The **Ok** button is disabled if the entered name:
    - Contains characters such as '*', '\', '/'.
    - Is just white-space.
    - Is one already being used.
- API test: unit tests for each of the public methods.

Case Study: Testing a 'Save as' Dialog

As with most unit tests, our test class has an `init()` method for getting an object into a known state, and a `cleanup()` method called at the end of each test. The instance variables are:

- A `JFrame` and a set of `IkonName`s from which the `SaveAsDialog` can be constructed.
- A `SaveAsDialog`, which is the object under test.
- A `UserStrings` and a `UISaveAsDialog` (listed later on) for manipulating the `SaveAsDialog` with keystrokes.
- A `ShowerThread`, which is a `Thread` for showing the `SaveAsDialog`. This is listed later on.

The outline of the unit test is:

```
public class SaveAsDialogTest {
    private JFrame frame;
    private SaveAsDialog sad;
    private IkonMakerUserStrings =
            IkonMakerUserStrings.instance();
    private SortedSet<IkonName> names;
    private UISaveAsDialog ui;
    private Shower shower;
    ...
    private void init() {
        ...
    }
    private void cleanup() {
        ...
    }
    private class ShowerThread extends Thread {
        ...
    }
}
```

UI Helper Methods

A lot of the work in this unit test will be done by the static methods in our helper class, `UI`. We looked at some of these (`isEnabled()`, `runInEventThread()`, and `findNamedComponent()`) in Chapter 8. The new methods are listed now, according to their function.

Dialogs

If a dialog is showing, we can search for a dialog by name, get its size, and read its title:

```java
public final class UI {
    ...
    /**
     * Safely read the showing state of the given window.
     */
    public static boolean isShowing( final Window window ) {
        final boolean[] resultHolder = new boolean[]{false};
        runInEventThread( new Runnable() {
            public void run() {
                resultHolder[0] = window.isShowing();
            }
        } );
        return resultHolder[0];
    }
    /**
     * The first found dialog that has the given name and
     * is showing (though the owning frame need not be showing).
     */
    public static Dialog findNamedDialog( String name ) {
        Frame[] allFrames = Frame.getFrames();
        for (Frame allFrame : allFrames) {
            Window[] subWindows = allFrame.getOwnedWindows();
            for (Window subWindow : subWindows) {
                if (subWindow instanceof Dialog) {
                    Dialog d = (Dialog) subWindow;
                    if (name.equals( d.getName() )
                                    && d.isShowing()) {
                        return (Dialog) subWindow;
                    }
                }
            }
        }
        return null;
    }
    /**
     * Safely read the size of the given component.
     */
    public static Dimension getSize( final Component component ) {
        final Dimension[] resultHolder = new Dimension[]{null};
```

```
            runInEventThread( new Runnable() {
                public void run() {
                    resultHolder[0] = component.getSize();
                }
            } );
            return resultHolder[0];
        }
        /**
         * Safely read the title of the given dialog.
         */
        public static String getTitle( final Dialog dialog ) {
            final String[] resultHolder = new String[]{null};
            runInEventThread( new Runnable() {
                public void run() {
                    resultHolder[0] = dialog.getTitle();
                }
            } );
            return resultHolder[0];
        }   ...
    }
```

Getting the Text of a Text Field

The method is `getText()`, and there is a variant to retrieve just the selected text:

```
//... from UI
/**
 * Safely read the text of the given text component.
 */
public static String getText( JTextComponent textComponent ) {
    return getTextImpl( textComponent, true );
}
/**
 * Safely read the selected text of the given text component.
 */
public static String getSelectedText(
                    JTextComponent textComponent ) {
    return getTextImpl( textComponent, false );
}
private static String getTextImpl(
        final JTextComponent textComponent,
        final boolean allText ) {
    final String[] resultHolder = new String[]{null};
```

```
        runInEventThread( new Runnable() {
            public void run() {
                resultHolder[0] = allText ? textComponent.getText() :
                        textComponent.getSelectedText();
            }
        } );
        return resultHolder[0];
}
```

Frame Disposal

In a lot of our unit tests, we will want to dispose of any dialogs or frames that are still showing at the end of a test. This method is brutal but effective:

```
//... from UI
public static void disposeOfAllFrames() {
    Runnable runnable = new Runnable() {
        public void run() {
            Frame[] allFrames = Frame.getFrames();
            for (Frame allFrame : allFrames) {
                allFrame.dispose();
            }
        }
    };
    runInEventThread( runnable );
}
```

Unit Test Infrastructure

Having seen the broad outline of the test class and the UI methods needed, we can look closely at the implementation of the test. We'll start with the UI Wrapper class and the `init()` and `cleanup()` methods.

The UISaveAsDialog Class

`UISaveAsDialog` has methods for entering a name and for accessing the dialog, buttons, and text field. The data entry methods use a `Cyborg`, while the component accessor methods use UI:

```
public class UISaveAsDialog {
    Cyborg robot = new Cyborg();
    private IkonMakerUserStrings us =
                IkonMakerUserStrings.instance();
```

Case Study: Testing a 'Save as' Dialog

```java
    protected Dialog namedDialog;
    public UISaveAsDialog() {
        namedDialog = UI.findNamedDialog(
                        SaveAsDialog.DIALOG_NAME );
        Waiting.waitFor( new Waiting.ItHappened() {
            public boolean itHappened() {
                return nameField().hasFocus();
            }
        }, 1000 );
    }
    public JButton okButton() {
        return (JButton) UI.findNamedComponent(
                IkonMakerUserStrings.OK );
    }
    public Dialog dialog() {
        return namedDialog;
    }
    public JButton cancelButton() {
        return (JButton) UI.findNamedComponent(
                IkonMakerUserStrings.CANCEL );
    }
    public JTextField nameField() {
        return (JTextField) UI.findNamedComponent(
                IkonMakerUserStrings.NAME );
    }
    public void saveAs( String newName ) {
        enterName( newName );
        robot.enter();
    }
    public void enterName( String newName ) {
        robot.selectAllText();
        robot.type( newName );
    }
    public void ok() {
        robot.altChar( us.mnemonic( IkonMakerUserStrings.OK ) );
    }
    public void cancel() {
        robot.altChar( us.mnemonic( IkonMakerUserStrings.CANCEL ) );
    }
}
```

A point to note here is the code in the constructor that waits for the name text field to have focus. This is necessary because the inner workings of Swing set the focus within a shown modal dialog as a separate event. That is, we can't assume that showing the dialog and setting the focus within it happen within a single atomic event. Apart from this wrinkle, all of the methods of `UISaveDialog` are straightforward applications of `UI` methods.

The ShowerThread Class

Since `SaveAsDialog.show()` blocks, we cannot call this from our main thread; instead we spawn a new thread. This thread could just be an anonymous inner class in the `init()` method:

```
private void init() {
    //Not really what we do...
    //setup...then launch a thread to show the dialog.
    //Start a thread to show the dialog (it is modal).
    new Thread( "SaveAsDialogShower" ) {
        public void run() {
            sad = new SaveAsDialog( frame, names );
            sad.show();
        }
    }.start();
    //Now wait for the dialog to show...
}
```

The problem with this approach is that it does not allow us to investigate the state of the `Thread` that called the `show()` method. We want to write tests that check that this thread is blocked while the dialog is showing.

Our solution is a simple inner class:

```
private class ShowerThread extends Thread {
    private boolean isAwakened;
    public ShowerThread() {
        super( "Shower" );
        setDaemon( true );
    }
    public void run() {
        Runnable runnable = new Runnable() {
            public void run() {
                sad.show();
            }
        };
```

```
            UI.runInEventThread( runnable );
            isAwakened = true;
        }
        public boolean isAwakened() {
            return Waiting.waitFor( new Waiting.ItHappened() {
                public boolean itHappened() {
                    return isAwakened;
                }
            }, 1000 );
        }
    }
```

The method of most interest here is `isAwakened()`, which waits for up to one second for the `awake` flag to have been set. This uses a class, `Waiting`, that is discussed in Chapter 12. We'll look at tests that use this `isAwakened()` method later on. Another point of interest is that we've given our new thread a name (by the call `super("Shower")` in the constructor). It's really useful to give each thread we create a name, for reasons that will be discussed in Chapter 20.

The init() Method

The job of the `init()` method is to create and show the `SaveAsDialog` instance so that it can be tested:

```
    private void init() {
        //Note 1
        names = new TreeSet<IkonName>();
        names.add( new IkonName( "Albus" ) );
        names.add( new IkonName( "Minerva" ) );
        names.add( new IkonName( "Severus" ) );
        names.add( new IkonName( "Alastair" ) );

        //Note 2
        Runnable creator = new Runnable() {
            public void run() {
                frame = new JFrame( "SaveAsDialogTest" );
                frame.setVisible( true );
                sad = new SaveAsDialog( frame, names );
            }
        };
        UI.runInEventThread( creator );
        //Note 3
        //Start a thread to show the dialog (it is modal).
        shower = new ShowerThread();
```

```
        shower.start();
        //Note 4
        //Wait for the dialog to be showing.
        Waiting.waitFor( new Waiting.ItHappened() {
            public boolean itHappened() {
                return UI.findNamedFrame(
                        SaveAsDialog.DIALOG_NAME ) != null;
            }
        }, 1000 );
        //Note 5
        ui = new UISaveAsDialog();
    }
```

Now let's look at some of the key points in this code.

Note 1: In this block of code we create a set of `IkonNames` with which our `SaveAsDialog` can be created.

Note 2: It's convenient to create and show the owning frame and create the `SaveAsDialog` in a single `Runnable`. An alternative would be to create and show the frame with a `UI` call and use the `Runnable` just for creating the `SaveAsDialog`.

Note 3: Here we start our `Shower`, which will call the blocking `show()` method of `SaveAsDialog` from the event thread.

Note 4: Having called `show()` via the event dispatch thread from our `Shower` thread, we need to wait for the dialog to actually be showing on the screen. The way we do this is to search for a dialog that is on the screen and has the correct name.

Note 5: Once the `SaveAsDialog` is showing, we can create our UI Wrapper for it.

The cleanup() Method

The `cleanup()` method closes all frames in a thread-safe manner:

```
    private void cleanup() {
        UI.disposeOfAllFrames();
    }
```

The Unit Tests

We've now done all the hard work of building an infrastructure that will make our tests very simple to write. Let's now look at these tests.

Case Study: Testing a 'Save as' Dialog

The Constructor Test

A freshly constructed `SaveAsDialog` should be in a known state, and we need to check the things we listed at the start of this chapter.

```
public boolean constructorTest() {
    //Note 1
    init();

    //Note 2
    //Check the title.
    assert UI.getTitle( ui.dialog() ).equals(
            us.label( IkonMakerUserStrings.SAVE_AS ) );

    //Note 3
    //Check the size.
    Dimension size = UI.getSize( ui.dialog() );
    assert size.width > 60;
    assert size.width < 260;
    assert size.height > 20;
    assert size.height < 200;

    //Note 4
    //Name field initially empty.
    assert UI.getText( ui.nameField() ).equals( "" );

    //Name field a sensible size.
    Dimension nameFieldSize = UI.getSize( ui.nameField() );
    assert nameFieldSize.width > 60;
    assert nameFieldSize.width < 260;
    assert nameFieldSize.height > 15;
    assert nameFieldSize.height < 25;

    //Ok not enabled.
    assert !UI.isEnabled( ui.okButton() );

    //Cancel enabled.
    assert UI.isEnabled( ui.cancelButton() );

    //Type in some text and check that the ok button is now enabled.
    ui.robot.type( "text" );
    assert UI.isEnabled( ui.okButton() );

    cleanup();
    return true;
}
```

Let's now look at the noteworthy parts of this code.

Note 1: In accordance with the rules for GrandTestAuto (see Chapter 19), the test is a public method, returns a `boolean`, and has name ending with "Test". As with all of the tests, we begin with the `init()` method that creates and shows the dialog. After the body of the test, we call `cleanup()` and return `true`. If problems are found, they cause an assert exception.

Note 2: The UI Wrapper gives us the dialog, and from this we can check the title. It is important that we are using the `UI` class to get the title of the dialog in a thread-safe manner.

Note 3: Here we are checking the size of dialog is reasonable. The actual allowed upper and lower bounds on height and width were found simply by trial and error.

Note 4: Although the UI Wrapper gives us access to the name field and the buttons, we must not interrogate them directly. Rather, we use our `UI` methods to investigate them in a thread-safe manner.

The wasCancelled() Test

The first of our API tests is to check the `wasCancelled()` method. We will basically do three investigations. The first test will call `wasCancelled()` before the dialog has been cancelled. The second test will cancel the dialog and then call the method. In the third test we will enter a name, cancel the dialog, and then call `wasCancelled()`.

There will be a subtlety in the test relating to the way that the **Cancel** button operates. The button is created in the constructor of the `SaveAsDialog`:

```
//From the constructor of SaveAsDialog
...
AbstractAction cancelAction = new AbstractAction() {
    public void actionPerformed( ActionEvent a ) {
        wasCancelled = true;
        dialog.dispose();
    }
};
JButton cancelButton = us.createJButton( cancelAction,
                        IkonMakerUserStrings.CANCEL );
buttonBox.add( Box.createHorizontalStrut( 5 ) );
buttonBox.add( cancelButton );
...
```

Case Study: Testing a 'Save as' Dialog

The action associated with the button sets the `wasCancelled` instance variable of the `SaveAsDialog`. The `wasCancelled()` method simply returns this variable:

```
From SaveAsDialog
public boolean wasCancelled() {
    return wasCancelled;
}
```

It follows that the `wasCancelled()` method is not thread-safe because the value it returns is set in the event thread. Therefore, in our test, we need to call this method from the event thread. To do this, we put a helper method into our test class:

```
//From SaveAsDialogTest
private boolean wasCancelled() {
    final boolean[] resultHolder = new boolean[1];
    UI.runInEventThread( new Runnable() {
        public void run() {
            resultHolder[0] = sad.wasCancelled();
        }
    } );
    return resultHolder[0];
}
```

Our `wasCancelledTest()` then is:

```
public boolean wasCancelledTest() {
    //When the ok button has been pressed.
    init();
    assert !wasCancelled();
    ui.saveAs( "remus" );
    assert !UI.isShowing( ui.dialog() );
    assert !wasCancelled();
    cleanup();

    //Cancel before a name has been entered.
    init();
    ui.cancel();
    assert !UI.isShowing( ui.dialog() );
    assert wasCancelled();
    cleanup();

    //Cancel after a name has been entered.
    init();
    ui.robot.type( "remus" );
    ui.cancel();
    assert !UI.isShowing( ui.dialog() );
    assert wasCancelled();
    cleanup();
    return true;
}
```

There are three code blocks in the test, corresponding to the cases discussed above, with each block of code being very simple and making use of the wasCancelled() helper method.

Writing code like the wasCancelled() method is pretty tiresome but is essential for solid tests. In fact, it's so important and so easily overlooked that we include it as a guideline:

[**Extreme Testing Guideline**: Any variable in a user interface or handler that is set from the event thread needs to be read in a thread-safe manner.]

Non-adherence to this guideline was the problem with our LoginScreenTest in Chapter 7. We checked the username and password passed back to the handler when **Ok** was pressed, but did not take care to do this safely, thereby guaranteeing intermittent test failures.

The name() Test

Like the wasCancelled() method, the name() method is not thread-safe, so our test class needs another boilerplate helper method:

```
//From SaveAsDialogTest
private IkonName enteredName() {
    final IkonName[] resultHolder = new IkonName[1];
    UI.runInEventThread( new Runnable() {
        public void run() {
            resultHolder[0] = sad.name();
        }
    } );
    return resultHolder[0];
}
```

Using this, we can write our nameTest():

```
public boolean nameTest() {
    init();
    //Note 1
    assert enteredName() == null;
    //Note 2
    ui.robot.type( "remus" );
    assert enteredName().equals( new IkonName( "remus" ) );
    //Note 3
    ui.ok();
    assert enteredName().equals( new IkonName( "remus" ) );
    cleanup();
    return true;
}
```

The main points of this test are as follows.

Note 1: Here we simply check that with no value entered into the text field, the method returns `null`. This could have gone into the constructor test.

Note 2: `UISaveAsDialog` has an `enterName()` method that types in the name and then presses *Enter*. In this test we want to type in a name, but not yet activate **Ok** by pressing *Enter*. So we use the `Cyborg` in the `UISaveDialog` to just type the name, and then we check the value of `name()`. This part of the test helps to define the method `SaveAsDialog.name()` by establishing that a value is returned even when the **Ok** button has not been activated.

Note 3: Here we are just testing that activating the **Ok** button has no effect on the value of `name()`. Later, we will also test whether the **Ok** button disposes the dialog.

It would be tempting to write some reflective method that made methods like `name()`, `wasCancelled()`, and `enteredName()` one-liners. However, that would make these examples much harder to understand. A bigger problem, though, is that we would lose compile-time checking: reflection breaks at runtime when we rename methods.

The show() Test

Our tests have used the `show()` method because it is used in `init()`. So we can be sure that `show()` actually does bring up the `SaveAsDialog` user interface. What we will check in `showTest()` is that the `show()` method blocks the calling thread.

```
public boolean showTest() {
    init();
    assert !shower.isAwakened();
    ui.cancel();
    assert shower.isAwakened();
    cleanup();

    init();
    assert !shower.isAwakened();
    ui.saveAs( "ikon" );
    assert shower.isAwakened();
    cleanup();
    return true;
}
```

In the first sub-test, we check that cancellation of the `SaveAsDialog` wakes the launching thread. In the second sub-test, we check that activation of the **Ok** button wakes the launching thread.

The Data Validation Test

The **Ok** button of the `SaveAsDialog` should only be enabled if the name that has been entered is valid. A name can be invalid if it contains an illegal character, or if it has already been used.

To test this behavior, we type in an invalid name, check that the **Ok** button is not enabled, then type in a valid name and test that it now is enabled:

```
ui.enterName( "*" );
assert !UI.isEnabled( ui.okButton() );
ui.enterName( "remus");
assert UI.isEnabled( ui.okButton() );
```

Our `validateDataTest()` started with a single block of code like that above. This block of code was copied and varied with different invalid strings:

```
public boolean validateDataTest() {
    //First check names that have illegal characters.
    init();
    ui.robot.type( "  " );
    assert !UI.isEnabled( ui.okButton() );
    ui.enterName( "remus");
    assert UI.isEnabled( ui.okButton() );

    ui.enterName( "*" );
    assert !UI.isEnabled( ui.okButton() );
    ui.enterName( "remus");
    assert UI.isEnabled( ui.okButton() );

    ...
    //Seven more blocks just like these.
    ...
    cleanup();
    return true;
}
```

Later on, this was refactored to:

```
public boolean validateDataTest() {
    init();
    //First check names that have illegal characters.
    checkOkButton( "  " );
    checkOkButton( "*" );
    checkOkButton( "/" );
    checkOkButton( "\\" );
    //Now names that are already there.
```

```
        checkOkButton( "albus" );
        checkOkButton( "Albus" );
        checkOkButton( "ALBUS" );
        checkOkButton( "MINERVA" );
        cleanup();
        return true;
    }
    private void checkOkButton( String name ) {
        ui.enterName( name );
        assert !UI.isEnabled( ui.okButton() );
        ui.enterName( "remus" );
        assert UI.isEnabled( ui.okButton() );
    }
```

The refactored code is much more readable, contains fewer lines, and does not contain the useless repeated test for the illegal string "*" that the original does. This illustrates a good point about writing test code. Because a lot of tests involve testing the state of an object against various simple inputs, it is very easy for such code to end up being unreadable and horribly repetitive. We should always be looking to refactor such code. Not only will this make the tests easier to maintain, it also makes the code more interesting to write. As we argued in Chapter 1, we should apply the same quality standards to our test code that we do to our production code.

The Usability Test

Typically, a simple dialog should be able to be cancelled with the *Escape* key, and the *Enter* key should activate the **Ok** button. In this test, we check these usability requirements and also check that tabbing to the buttons and activating them with the space key works as expected.

```
    public boolean usabilityTest() {
        //Check that 'escape' cancels.
        init();
        ui.robot.escape();
        assert !UI.isShowing( ui.dialog() );
        assert wasCancelled();
        cleanup();

        //Check activating the cancel button when it has focus.
        init();
        ui.robot.tab();//Only one tab needed as ok is not enabled.
        ui.robot.activateFocussedButton();
        assert !UI.isShowing( ui.dialog() );
        assert wasCancelled();
        cleanup();
```

```
        //Check that 'enter' is like 'ok'.
        init();
        ui.robot.type( "remus" );
        ui.robot.enter();
        assert !UI.isShowing( ui.dialog() );
        assert !wasCancelled();
        assert enteredName().equals( new IkonName( "remus" ) );
        cleanup();

        //Check activating the ok button when it has focus.
        init();
        ui.robot.type( "remus" );
        ui.robot.tab();
        ui.robot.activateFocussedButton();
        assert !UI.isShowing( ui.dialog() );
        assert !wasCancelled();
        assert enteredName().equals( new IkonName( "remus" ) );
        cleanup();
        return true;
    }
```

Summary

This chapter and the two preceding ones have given us all the principles we need to write solid, automated, and fairly painless tests for our user interfaces. The key points are:

- We should write a UI Wrapper class for the class we are testing and use it to manipulate the test objects.
- All creation and setup of components must be done in the event thread.
- All querying of the state of components must be done in a thread-safe manner.
- Any variable, either in a user interface or in a handler, that is set from the event thread needs to be read in a thread-safe manner.

The class UI contains a lot of methods for making it easy to follow these principles. In the next chapter we'll see more of this useful class.

10
More Techniques for Testing Swing Components

In the last several chapters, we have developed a lot of techniques for writing user interface tests. For mimicking user actions we have used `Cyborg`, usually via a UI Wrapper class. For checking the state of components, we've used our `UI` class. These tests have been limited to user interfaces involving text fields and various kinds of buttons. In this chapter, we'll see how to test more complex user interface elements such as color choosers, file choosers, tables, trees, sliders, progress bars, menus, and active drawing areas. Generally speaking, there are interrogation methods in `UI` and manipulation methods in `Cyborg`. Together, these classes provide most of the tools we need for automated user interface testing.

In this chapter, we'll continue to use 'Ikon Do It' as our source of examples.

Testing with JColorChooser

Swing provides a ready-made component for selecting colors — `JColorChooser`. By default this features three tabs, each providing a different method for selecting a color. When a `JColorChooser` is shown, the **Swatches** tab is selected.

The following figure shows the default appearance of a color chooser, for an English-speaking locale:

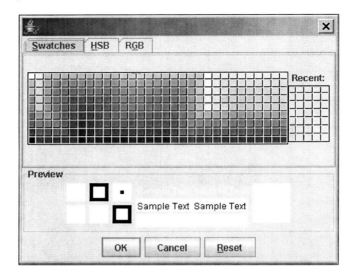

Suppose that in one of our tests we want to select a specific color. This can be achieved as follows:

- Get the `JColorChooser` displayed, using some means specific to our software.
- Select the **RGB** tab, so that we can enter the red, green, and blue components of the color.

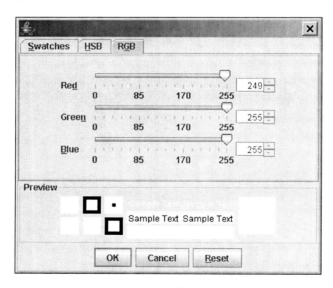

- To enter the red component of our color:
    ◦ Use the *Alt+d* mnemonic to get focus to the red slider.
    ◦ Then use *Tab* to get focus to the spinner.
    ◦ Press *Delete* three times to delete whatever numbers are in the field. There is no harm in pressing delete more times than there are digits in the field, and there are never more than three digits.
    ◦ Enter the red component.
    ◦ Use *Enter* to lock it in.
- Enter the green and blue components in a similar way.
- Activate the **Ok** button to finalize the color selection.

The only tricky thing here is that the mnemonics mentioned in the procedure above will not work for languages other than English. To get the mnemonics for the locale in which our program is running, we use javax.swing.UIManager. The mnemonic for the **RGB** swatch is obtained using the key ColorChooser.rgbMnemonic, the key for the red field mnemonic is ColorChooser.rgbRedMnemonic, and so on. We can see that these keys are used by looking at the source code for DefaultRGBChooserPanel in javax.swing.colorchooser. Once we know these keys, we can implement our procedure for using JColorChooser in a locale-neutral fashion.

```
//From the unit tests for IkonMaker, a component of Ikon Do It.
public class UIIkonMaker {
    private Cyborg cyborg = new Cyborg();
    ...
    public void selectColour( Color colour ) {
        //Choose the RGB swatch.
        cyborg.altChar( mn("ColorChooser.rgbMnemonic") );
        enterColourComponent( mn("ColorChooser.rgbRedMnemonic"),
            colour.getRed() ) ;
        enterColourComponent( mn("ColorChooser.rgbGreenMnemonic"),
            colour.getGreen() ) ;
        enterColourComponent( mn("ColorChooser.rgbBlueMnemonic"),
            colour.getBlue() ) ;
        //Return to the colour tiles swatch,
        //which does not use any mnemonics
        //used elsewhere.
        cyborg.altChar( mn("ColorChooser.swatchesMnemonic") );
    }
    private int mn( String uiManagerConstant ) {
        Object uiValue = UIManager.get(uiManagerConstant);
        return Integer.parseInt( uiValue.toString() );
```

```
        }
        private void enterColourComponent( int mnemonic, int value ) {
            cyborg.altChar( mnemonic );
            cyborg.tab();//Go to text field
            cyborg.selectAllText();
            cyborg.type( "" + value );
            cyborg.enter();
        }
    }
```

The implementation above is from the UI wrapper for `IkonMaker`, which is a component of 'Ikon Do It' application. `IkonMaker` has a `JColorChooser` embedded in it, so there is no need to show the color chooser dialog. However, there is the added wrinkle that we must return to the color tiles swatch after selecting our color, because the **RGB** swatch shares mnemonics with other parts of `IkonMaker`.

Using JFileChooser

Using `JFileChooser` is even easier than using a `JColorChooser`. The default behavior for a `JFileChooser` is to show a text field that already has the focus and into which the full path of the file to be selected can be typed. Also, the accept button is the default button, so pressing *Enter* activates it. (The label of the accept button depends on how we set up the `JFileChooser`.) Assuming that we have activated our file chooser in some software-specific fashion, it is then easy to select a file. For example:

```
    public void useImage( File imageFile ) {
        invokeUseImage(); //Application specific.
        cyborg.type( imageFile.getAbsolutePath() );
        cyborg.enter();
    }
```

Checking that a JFileChooser has been Set Up Correctly

With ready-made components such as `JFileChooser` and `JColorChooser`, we can generally assume that if we have set them up correctly, they will work correctly. However, in setting up these components there are always plenty of opportunities for making mistakes. Here are some properties of a `JFileChooser` that we might need to set:

- The `FileFilter`, which determines which files will show.
- The description of these files.

- Whether files, directories, or both files and directories can be selected.
- Whether multiple files can be selected.
- The initial selection.
- The text on the accept button.

How can we test that these properties have been correctly set?

Using our UI class and its powerful search mechanism, this is easy. We simply show the JFileChooser, find it, and then check its properties. Here is how the UI class finds a JFileChooser:

```
public static JFileChooser findFileChooserThatIsCurrentlyShowing() {
    return (JFileChooser) findComponentInSomeFrame(
            new ComponentSearchCriterion() {
        public boolean isSatisfied( Component component ) {
            return (component instanceof JFileChooser)
                    && component.isShowing();
        }
    } );
}
```

The code below is a typical test (of IkonMaker) that finds a JFileChooser and checks some of its properties. In this code, ui is a wrapper for the IkonMaker main screen. The test creates an icon and then invokes a file chooser to select an image to use in the icon. We check that the file chooser shows image files and allows directory navigation. Then we check that if invoked again, the file chooser is showing the directory from which our previous image was taken.

```
public boolean useImageFileChooserTest() {
    init();//Creates and shows the IkonMaker.
    ui.createNewIkon( "ike1", 16, 16 );
    ui.invokeUseImage();
    JFileChooser chooser =
            UI.findFileChooserThatIsCurrentlyShowing();
    //Check that directories and files can be selected
    //(directories so that one can navigate
    //around the file system).
    assert chooser.getFileSelectionMode() ==
            JFileChooser.FILES_AND_DIRECTORIES;
    FileFilter ff = chooser.getFileFilter();
    //Check that .png, .gif, .jpg, and .jpeg
    //files are accepted, and that
    //case is ignored in checking file extensions.
    File png = new File( testDataDir, "RGGR.png" );
    assert ff.accept( png );
    File gif = new File( testDataDir, "BRRB.gif" );
    assert ff.accept( gif );
```

More Techniques for Testing Swing Components

```
        File jpg = new File( testDataDir, "waratah.jpg" );
        assert ff.accept( jpg );
        File jpeg = new File( testDataDir, "BirdInBush.JPEG" );
        assert ff.accept( jpeg );
        //Actually choose the png image. Then re-invoke the
        //file chooser and check that it is looking at the
        //directory containing the png image.
        cyborg.type( png.getAbsolutePath() );
        cyborg.enter();
        ui.invokeUseImage();
        chooser = UI.findFileChooserThatIsCurrentlyShowing();
        File currentDir = chooser.getCurrentDirectory();
        assert currentDir.equals( testDataDir ) :
                "Current dir: " + currentDir;
        cyborg.escape();//Close the chooser.
        cleanup();
        return true;
    }
```

Testing the Appearance of a JComponent

'Ikon Do It' contains two `IkonCanvas` objects: one is a large and active drawing screen and the other is a preview screen. Here is a partially complete, very bad drawing of a cat (no real cat was hurt in the production of this book):

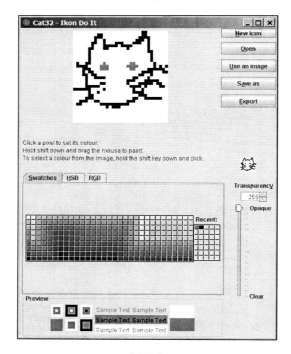

The active canvas is listening for mouse events (as is the cat, no doubt) and updating an underlying `Ikon` object. The unit tests for `IkonCanvas` need to check that when the color of a pixel is changed in the model, either programmatically or by mouse events, the canvas gets re-drawn correctly. To write this test, we need to be able to check the appearance of a component, pixel-by-pixel.

To put this in context, here is part of the unit test for `IkonCanvas`:

```
public boolean mouseDraggedTest() {
    //First we create an IkonCanvas backed by an 8-by-8
    //Ikon in which all pixels are blue. The color to paint
    //with is red.
    guiInit( 8, 8, Color.BLUE );
```

After this initialization, the `IkonCanvas` appears like this with the blue canvas:

```
    //Drag from top left to bottom right.
    ui.dragMouse( 0, 0, 7, 7 );
```

After the mouse dragging, the canvas appears like shown in the next figure:

Now we need to check that the component really is drawn as we expect. We set up an array of expected pixel colors.

```
Color[][] expected = new Color[8][8];
for (int i = 0; i < expected.length; i++) {
    for (int j = 0; j < expected[i].length; j++) {
        if (i != j) {
            expected[i][j] = Color.BLUE;
        } else {
            expected[i][j] = colourToSupply;//i.e. red
        }
    }
}
//Check that the underlying Ikon has been
```

```
//correctly updated by the mouse dragging.
checkIkonColours( expected );

//Now check that the component is correctly painted.
checkIkonCanvas( canvas );
```

It is the `checkIkonCanvas()` method that we are interested in here.

```
void checkIkonCanvas( IkonCanvas ic ) {
    //For each pixel in the ikon, check that the square
    //representing it in the canvas is correctly colored.
    //An IkonCanvas has a JComponent, the location of which
    //we find in a thread-safe way.
    Point screenLocation = UI.getScreenLocation( ic.component() );

    //The data model for the IkonCanvas is an Ikon object,
    //which has an array of pixel data.
    //Loop through the data in the IkonCanvas...
    for (int ikonRow = 0; ikonRow < ic.ikon().height(); ikonRow++) {
        for (int ikonCol = 0;
             ikonCol < ic.ikon().width(); ikonCol++) {

            //This is the color that the pixel is in the data model.
            Color pixelColour =
                    ic.ikon().colourAt( ikonRow, ikonCol );

            //An IkonCanvas represents an Ikon pixel
            //as a square, the sides of which are
            //available as pixelSize().
            //The on-screen location of the square representing the
            //data pixel needs to be calculated...
            int topLeftX = screenLocation.x +
                                ikonCol * ic.pixelSize();
            int topLeftY = screenLocation.y +
                                ikonRow * ic.pixelSize();

            //Now we can check each screen pixel
            //in the little square representing the data pixel.
            //Check that each pixel in the part of the ikon canvas
            //component that is representing this pixel
            //is coloured correctly.
            for (int repRow = 0; repRow < ic.pixelSize();
                                                    repRow++) {
                for (int repCol = 0;
                     repCol < ic.pixelSize(); repCol++) {
                    //The actual check of a screen pixel
                    //is done by Cyborg.
                    cyborg.checkPixel( topLeftX + repCol,
```

```
                              topLeftY + repRow, pixelColour );
                    }
                }
            }
        }
    }
```

The point of this example is that checking the appearance of a component, which is often regarded as too hard to be worth trying to do, is actually pretty easy using `Cyborg`. This is true as long as we know exactly what the component should look like, which is generally the case for animations and drawings.

Testing with Frames

When an application opens, its main frame or dialog should be wherever it was left when the application was last shut down. Frame size, too, should be maintained between uses of the application.

Frame Location

Below is the `IkonMaker` unit test for frame location. The instance variable `ui` is an `UIIkonMaker`. As usual, the `init()` method creates and shows the `IkonMaker`, and then creates the `ui` reference.

```
    /**
     * Checks that the location of the main frame is written to
     * and read from user preferences.
     */
    public boolean frameLocationTest() {
        init();
        //Note 1
        //Get the current location.
        Point initialLocation = UI.getScreenLocation( ui.frame() );

        //Note 2
        //Move the frame a bit.
        cyborg.dragFrame( ui.frame(), 100, 200 );
        initialLocation.translate( 100, 200 );

        //Note 3
        Point newLocation = UI.getScreenLocation( ui.frame() );
        //Sanity check.
        assert newLocation.equals( initialLocation ) :
                "Expected: " + initialLocation + ", got: " +
                                                newLocation;
```

```
        //Note 4
        //Close the IkonMaker.
        ui.exit();
        //Check that the new location is stored in preferences.
        int prefsX = preferences.getInt(
                    IkonMaker.PREFERENCES_KEY_LOCATION_X,
                                    Integer.MIN_VALUE );
        int prefsY = preferences.getInt(
                    IkonMaker.PREFERENCES_KEY_LOCATION_Y,
                                    Integer.MIN_VALUE );
        assert newLocation.x == prefsX;
        assert newLocation.y == prefsY;
        //Now build a new IkonMaker.
        new IkonMaker( storeDir );
        ui = new UIIkonMaker();
        //Check that it's in the right spot.
        newLocation = UI.getScreenLocation( ui.frame() );
        assert newLocation.equals( initialLocation ) :
                    "Expected: " + initialLocation + ", got: "
                                            + newLocation;
        cleanup();
        return true;
    }
```

Let's now look at the key points of this test.

Note 1: We're not testing what the initial location of the frame is. We just need to record it in order to check that it is changed by the frame dragging that occurs later in the test.

Note 2: This test makes use of the `Cyborg` method for moving a frame, which works by dragging the frame with the mouse:

```
    //From Cyborg.
    /**
     * Move the given frame by dragging with the mouse.
     * @param frame the frame to be moved
     * @param deltaX the horizontal displacement
     * @param deltaY the vertical displacement
     */
    public void dragFrame( Frame frame, int deltaX, int deltaY ) {
        Point initialLocation = UI.getScreenLocation( frame );
        Point destination = new Point( initialLocation );
        destination.translate( deltaX,   deltaY );
        //From trial and error, we know that clicking
        //40 pixels to the left, and 5 down, from the top
        //left corner of a frame allows us to drag it.
        //Adjust the locations accordingly.
        initialLocation.translate( 40, 5 );
```

```
        destination.translate( 40, 5 );
        mouseDrag( initialLocation, destination );
}
```

Note 3: Here we are checking that the test setup works — a 'sanity check'. Although the `dragFrame()` method is tested and can be assumed to work, this kind of check can be reassuring and reduces the scope of the search for errors should the test fail.

Note 4: Here `preferences` is the `java.util.Preferences` instance used by the `IkonMaker`. It is globally available and refreshed in the `init()` method:

```
//Get the preferences for the IkonMaker and clear them out.
preferences = Preferences.userNodeForPackage( IkonMaker.class );
try {
    preferences.clear();
} catch (BackingStoreException e) {
    e.printStackTrace();
    assert false : "Could not clear preferences, as shown.";
}
```

The use of this `preferences` object and the `preferences` keys make this test sensitive to internal changes to the code of `IkonMaker`, as is generally the case for "white box" tests. It is a convenient strategy for this test as then we do not have to check that the user's preference for frame location is preserved between JVMs. The test has shown (beyond reasonable doubt) that the frame location is stored using `Preferences`. When we build the new `IkonMaker`, and show that it is in the correct location, we prove (again, beyond reasonable doubt) that the frame is located using the values read from the `preferences` system. These facts, and the assumption that the `preferences` system works between JVMs (which we can take as given) convince us that the user's frame preferences are persisted between runs of the program.

Frame Size

An almost identical test to the one above can be used to show that the user's preferences for frame size are respected. The key ingredient in this test is a `Cyborg` method for re-sizing a frame:

```
//From Cyborg.
/**
 * Re-size the given frame by dragging the lower
 * right corner of it with the mouse.
 * @param frame the frame to be resized.
 * @param deltaWidth the increase in width
 * @param deltaHeight the increase in height.
 */
public void resizeFrame( Frame frame,
                    int deltaWidth, int deltaHeight ) {
```

```
        Point location = UI.getScreenLocation( frame );
        Dimension initialSize = UI.getSize( frame );
        //Click just inside the bottom right corner.
        Point placeToClick = new Point(
                location.x + initialSize.width - 2,
                location.y + initialSize.height - 2);
        Point placeToDragTo = new Point( placeToClick.x + deltaWidth,
                placeToClick.y + deltaHeight );
        mouseDrag( placeToClick,  placeToDragTo );
}
```

Testing with Lists

Lists are commonly used components and there are `Cyborg` methods for operating them and `UI` methods for investigating their state.

List Selection Methods

If a list has focus then the *Home* key selects the first element and the *End* key selects the last. (The complete list of these key actions can be found at http://java.sun.com/products/jlf/ed2/book/Appendix.A8.html.) These keystrokes are used to implement `Cyborg` methods `selectFirstOnlyInList()` and `selectLastOnlyInList()`. More complex selections are possible too:

```
//From Cyborg.
/**
 * Assuming that a list has the focus, selects the given indices,
 * and deselects any others that are currently selected.
 *
 * @param the indices to be selected, in increasing order.
 */
public void selectListIndicesOnly(
        final int[] toSelect, int listSize ) {
    //Select just the first element. We will undo below if need be.
    selectFirstOnlyInList();
    //An array of flags representing the indices to be selected.
    boolean[] toBeSelected = new boolean[listSize];
    for (int i = 0; i < toSelect.length; i++) {
        toBeSelected[toSelect[i]] = true;
    }
    //Now do the selections....
    //Press the control button.
    robot.keyPress( KeyEvent.VK_CONTROL  );
    //Do we need to de-select the first item?
    if (!toBeSelected[0]) space();
    //Go through the remaining items.
    for (int i=1; i<listSize; i++) {
```

```
            //Move to the item.
            down();
            //Select if necessary.
            if (toBeSelected[i]) space();
        }
        //Release the control button.
        robot.keyRelease( KeyEvent.VK_CONTROL );
    }
```

There is also a UI method for reading the selected indices of a list:

```
From UI.
/**
 * Safely read the selected indices in the given list.
 */
public static int[] getSelectedIndices( final JList list ) {
    final int[][] resultHolder = new int[1][];
    runInEventThread( new Runnable() {
        public void run() {
            resultHolder[0] = list.getSelectedIndices();
        }
    } );
    return resultHolder[0];
}
```

List Rendering

The 'Ikon Do It' dialog for opening an icon presents a list of the icons. The following figure shows a list with specialized rendering from the 'Ikon Do It' application:

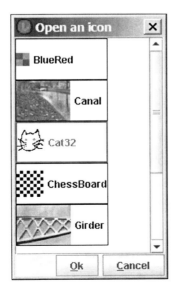

When the elements of a list are complex like this, we want to test that they are rendered correctly. This can be done as follows. From a `JList` we can get the `ListCellRenderer` that is used to render each entry of the list. (Each `JList` has a `ListCellRenderer` that is used to draw components for the list entries. These components are printed onto the screen to represent the list data, but are not actually laid out.) We can then investigate the kinds of `Component` produced by the renderer.

For example, the `ListCellRenderer` for the icon list shown previously produces components with a border that have color and thickness that depend on whether the corresponding list item is selected. This can be verified as follows:

```java
public boolean listRendererTest() {
    init();
    //Get the list object and from it the renderer.
    final JList list = ui.namesList();
    //Have to get the renderer in the event thread.
    final ListCellRenderer[] lcrHolder = new ListCellRenderer[1];
    UI.runInEventThread( new Runnable() {
        public void run() {
            lcrHolder[0] = list.getCellRenderer();
        }
    } );
    final ListCellRenderer lcr = lcrHolder[0];

    //An ikon to be rendered.
    String name = "IkonName";
    final Ikon ike = buildIkon( name, false );

    //If not selected, the border should be
    //black and one pixel wide.
    //Have to get the component in the event thread.
    final Component[] componentHolder = new Component[1];
    UI.runInEventThread( new Runnable() {
        public void run() {
            componentHolder[0] =
                    lcr.getListCellRendererComponent(
                        list, ike, 0, false, false );
        }
    } );
    Box box = (Box) componentHolder[0];
    LineBorder border = (LineBorder) box.getBorder();
    assert border.getLineColor().equals( Color.BLACK );
    assert border.getThickness() == 1;

    cleanup();
    return true;
}
```

This test is assuming a lot about the internals of the class it is testing. For example, that the component is a `Box` with a `LineBorder`. As with the frame location test, this use of "inside knowledge" makes this "white box" test sensitive to changes in implementation of the class itself. However, since we are testing such a fine detail of the class, this is not necessarily a big problem, especially if we are running our tests as part of a continuous build process.

List Properties

It is easy to make mistakes in setting up a `JList`, because it is so highly customizable. Some things to check are:

- Selection mode.
- Layout orientation.
- Visible row count.

These can be tested by direct interrogation of the list object. For example, from the unit test for the open icon dialog:

```
//Check that the list allows only single selection.
assert ui.namesList().getSelectionMode()
             == ListSelectionModel.SINGLE_SELECTION;
```

Testing a JTable

Some things that we might do in a unit test for a component that has a `JTable` are to right or left-click a particular cell, click a column header (to re-order, perhaps), or permute the columns.

`Cyborg` and `UI` together provide the means to do these things. The `UI` class gives methods for locating table and table header cells. To click a cell, `Cyborg` clicks five pixels down and to the right of the top left hand corner of the cell. This will fail for tables with tiny cells, but we have never found this to be a problem in practice.

```
//From Cyborg.
/**
 * Click just inside the indexed cell of the given table.
 */
public void clickTableCell( JTable table, int row, int column) {
    mouseLeftClickInsideComponentAt(
            UI.getCellPositionOnScreen( table, row, column ) );
}
/**
```

```java
 * Click just inside the indexed header cell of the given table.
 */
public void clickTableHeader( JTable table, int column ) {
    mouseLeftClickInsideComponentAt(
                UI.getColumnHeaderPosition( table, column ) );
}
/**
 * Drag the indexed column to another position.
 */
public void dragTableColumn(
        JTable table, int column, int destination ) {
    Point from = pointJustInsideComponentAt(
                UI.getColumnHeaderPosition( table, column ) );
    Point to = pointJustInsideComponentAt(
                UI.getColumnHeaderPosition( table, destination ) );
    mouseDrag( from, to );
}
private void mouseLeftClickInsideComponentAt( Point position ) {
    mouseLeftClick( pointJustInsideComponentAt( position ) );
}
private Point pointJustInsideComponentAt( Point position ) {
    return new Point( position.x + 5, position.y + 5 );
}
```

The unit tests for these methods are more interesting than the methods themselves. For example, the test for UI.getCellPositionOnScreen(JTable, int, int) creates a JTable with all cells having the same size. Then, using the table location, it is possible to calculate the location of each cell and compare it with that returned by getCellPositionOnScreen():

```java
//From UITest.
public boolean getCellPositionOnScreenTest() {
    //Create and show a table with 7 columns and 4 rows.
    //Define the cell renderer so that each cell is
    //80 by 80.
    final JTable[] tables = setUpTable();

    //Using the table location, and known cell dimensions,
    //calculate the location of each cell, and check
    //against that given by the method.
    Point tableLocation = UI.getScreenLocation( tables[0] );
    for (int row = 0; row < 4; row++) {
        for (int col = 0; col < 7; col++) {
            int expectedCellX = col * tableCellSize.width
                                    + tableLocation.x;
```

```
                int expectedCellY = row * tableCellSize.height
                                            + tableLocation.y;
                Point expectedLocation =
                        new Point( expectedCellX, expectedCellY );
                Point actualCellPosition =
                     UI.getCellPositionOnScreen( tables[0], row, col );
                assert actualCellPosition.equals( expectedLocation ) :
                              "Expected: " + expectedLocation +
                                    ", got: " + actualCellPosition;
            }
        }
        cleanup();
        return true;
    }
```

The test for `Cyborg.dragTableColumn(JTable, int, int)` creates a table to which is attached a `TableColumnModelListener` that will record column re-arrangements:

```
    //From CyborgTest.
    private class RecordingTableColumnModelListener
                        implements TableColumnModelListener {
        List< Integer>  fromColumns = new LinkedList<Integer>( );
        List< Integer>  toColumns = new LinkedList<Integer>( );
        public void columnMoved( TableColumnModelEvent e ) {
            //There are heaps of events generated by a column drag,
            //we ignore those that are from a column to itself.
            if (e.getFromIndex() != e.getToIndex()) {
                fromColumns.add( e.getFromIndex() );
                toColumns.add( e.getToIndex() );
            }
        }
        public void columnAdded( TableColumnModelEvent e ) {}
        public void columnRemoved( TableColumnModelEvent e ) {}
        public void columnMarginChanged( ChangeEvent e ) {}
        public void columnSelectionChanged( ListSelectionEvent e ) {}
    }
```

The test itself is then pretty straightforward:

```
    //From CyborgTest.
    public boolean dragTableColumnTest() {
        initWithTable();
        borg.dragTableColumn( table,1, 4);
        List<Integer> expectedFromColumns
                        = new LinkedList<Integer>();
        expectedFromColumns.add( 1 );
        expectedFromColumns.add( 2 );
        expectedFromColumns.add( 3 );
        List<Integer> expectedToColumns
                = new LinkedList<Integer>();
```

```
            expectedToColumns.add( 2 );
            expectedToColumns.add( 3 );
            expectedToColumns.add( 4 );
            assert tcml.fromColumns.equals( expectedFromColumns );
            assert tcml.toColumns.equals( expectedToColumns );
            cleanup();
            return true;
        }
```

The rendering of table cells can be checked in the same way as that of `JList`. Similarly, the properties of the table (for example, is row selection allowed) can be checked directly.

Testing with JMenus

As with lists and tables, there are methods in `UI` for checking `JMenus`, and methods in `Cyborg` for manipulating them. We'll start with the `UI` methods.

Checking the Items

A `JMenu` can be tested like any other user interface component. Some basic things that we might want to check are:

- The correct items are present.
- The items are enabled and disabled correctly according to the state of the user interface.
- Each item has a mnemonic.
- No mnemonic is used twice.

These last two requirements might not be appropriate for very long menus, but should be followed as far as possible.

We also need to test that our menu items do what they say. For example, that **File | Exit** actually closes an application. The (not very scientific) method of doing this is to write tests not only for each user action, but also for each different way of invoking that action. For now, let's look at how to implement the 'cosmetic' tests listed above.

```
    //From UI.
    /**
     * Finds the named menu and checks that the n-th item
     * has text <code>us.label( itemKeys[n] )</code> and
     * enabled status equal to <code>enabledFlags[n]</code>.
     * Also checks that every item has a mnemonic, that
     * no mnemonic is used twice, and that there
     * are no more or less items than expected.
```

```
 * Any failure will result in an assertion error
 * with a suitable message.
 */
//Note 1
public static void checkMenu( final String name,
                              final UserStrings us,
                              final String[] itemKeys,
                              final boolean[] enabledFlags ) {
    //Note 2
    final String[] problemsFound = new String[1];
    problemsFound[0] = null;
    Runnable problemFinder = new Runnable() {
        public void run() {
            //Get the menu itself.
            //Note 3
            JMenu menu = (JMenu) findNamedComponent( name );
            //Check the components..
            checkJMenuItems( menu.getMenuComponents(),
                    enabledFlags, problemsFound, us, itemKeys );
        }
    };
    runInEventThread( problemFinder );
    if (problemsFound[0] != null) {
        assert false : problemsFound[0];
    }
}
```

Let's look at the main points of interest in this test.

Note 1: This method will only work if the JMenu to be checked has been given a name. Also, the labels and mnemonics in the menu must have been set using the UserStrings pattern described in Chapter 5. The fact that menus created using the UserStrings pattern can be thoroughly tested is another good reason for using the pattern.

Note 2: The actual checking will require a lot of investigation of component states, which of course must be done in the event thread. The problemsFound array is for communication of errors from the event thread to the calling thread.

Note 3: Given the name of the menu, we can easily find it and then traverse over the components. The actual checking is delegated to a private method that is also used in checkPopupMenu(). Here is this private method:

```
//From UI.
private static void checkJMenuItems( Component[] components,
                                     boolean[] enabledFlags,
                                     String[] problemsFound,
                                     UserStrings us,
```

```java
                            String[] itemKeys ) {
    //Iterate over the components. For those that are JMenuItems,
    //check the text, the enabled state, and the mnemonic.
    int index = 0;
    //Note 4
    Set<Integer> mnemonicsUsed = new HashSet<Integer>();
    for (int i = 0; i < components.length; i++) {
        Component component = components[i];
        //Note 5
        if (component instanceof JMenuItem) {
            JMenuItem item = (JMenuItem) component;
            //Check that there are not more
            //components than expected.
            if (index >= enabledFlags.length) {
                problemsFound[0] = "" +
                        "More components than expected.";
                break;
            }
            //Check the text.
            String expectedText = us.label( itemKeys[index] );
            if (!expectedText.equals( item.getText() )) {
                problemsFound[0] = "At index " + index +
                        " expected " + expectedText +
                        " but got " + item.getText();
                break;
            }
            //Check the enabled state.
            if (item.isEnabled() != enabledFlags[index]) {
                problemsFound[0] = "At index " + index +
                        " enabled state was: "
                        + item.isEnabled();
                break;
            }
            //Check that there is a mnemonic.
            int mnemonic = item.getMnemonic();
            if (mnemonic == 0) {
                problemsFound[0] =
                        "No mnemonic at index " + index;
                break;
            }
            //Check that the mnemonic
            //has not already been used.
            if (mnemonicsUsed.contains( mnemonic )) {
                problemsFound[0] =
                        "Mnemonic " + mnemonic +
                        " re-used at index " + index;
                break;
```

```
                    }
                    mnemonicsUsed.add( mnemonic );
                    //Increment index.
                    index++;
                }
            }
            //Check that there are not too few components.
            //Skip this if problems already found, in which
            //case index will be less than the number
            //of components.
            if (problemsFound[0] == null
                    && index < enabledFlags.length) {
                problemsFound[0] =
                        "Wrong number of components: " + index;
            }
        }
```

Note 4: Recording the mnemonics as we find them allows us to check that no mnemonic is used twice.

Note 5: We skip over components that are not `JMenuItems`, such as separators. This block first checks that we have not exceeded the expected number of items and then checks on a particular menu item. We exit as soon as we find a problem.

Examples of how to use this method in practice can be found in the unit tests for `UI`.

Using Menus with Cyborg

Assuming that a menu has been built using labels and mnemonics from a `UserStrings`, `Cyborg` provides a convenient way of activating an item from the menu:

```
From Cyborg.
/**
 * Activates the menu using the mnemonic
 * <code>us.mnemonic( menuKey )</code> and then
 * activates the menu item with mnemonic
 * <code>us.mnemonic( itemKey )</code>.
 */
public void activateMenuItem(
        UserStrings us,
        String menuKey, String itemKey ) {
    altChar( us.mnemonic( menuKey ));
    type( (char) us.mnemonic( itemKey).intValue() );
}
```

Testing JPopupMenus

Contextual popup menus are great for making programs easy to use. These menus are generally invoked with a right mouse click or, in the Windows world, by a dedicated key, which can be invoked using `Cyborg.contextMenu()`. (This chapter was originally written on an old laptop with a broken touchpad, on a train. Plugging in a mouse was not an option, as the next passenger's lap did not seem to be good mouse mat material. It is amazing how efficiently one can work without a mouse when forced to. The 'context' key is essential for this. The lack of support for this key in some Java programs has been a real annoyance.)

Because of their contextual nature, we need an easy way to check the contents of popup menus. `UI` provides this method as `checkPopup()`, which is very similar to `checkMenu()` discussed in the previous section.

Toolbars can be tested in much the same way as popups and other menus.

Combo Boxes

Combo boxes are very commonly used components and we provide methods in `Cyborg` for manipulating them with keyboard events, and methods in `UI` for reading their state.

The most interesting of these is the `Cyborg` method to select an indexed element of the currently focused combo:

```
//From Cyborg.
/**
 * Assuming that a combo box is in focus,
 * selects the indexed element.
 */
public void selectElementInComboByIndex( int index ) {
    selectFirstInCombo();//So we know where we are.
    space();//Brings up the list view.
    for (int i=0; i<index; i++) down();
    enter();//Lock it in!
}
```

This method assumes that the `JComboBox` to be manipulated is the focused component. How can we get focus to our `JComboBox`? Well, generally an active component such as a combo has an associated label, and the label mnemonic can be used to give focus to the component. As an example, consider this user interface mocked up in one of the `Cyborg` unit tests:

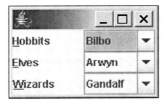

In the preceding figure, the combo boxes are activated by the mnemonics for the corresponding labels.

The unit test operates the combos as follows:

```
//From CyborgTest.
//Here, borg is a Cyborg.
borg.altChar( 'E' );//Focus to the elves.
borg.selectElementInComboByIndex( 1 );
assert UI.getSelectedComboIndex( combos[0] ) == 0;
assert UI.getSelectedComboIndex( combos[1] ) == 1;
assert UI.getSelectedComboIndex( combos[2] ) == 0;
borg.altChar( 'H' );//Focus to the hobbits.
borg.selectElementInComboByIndex( 2 );
assert UI.getSelectedComboIndex( combos[0] ) == 2;
assert UI.getSelectedComboIndex( combos[1] ) == 1;
assert UI.getSelectedComboIndex( combos[2] ) == 0;
...
```

Progress Bars

When our software is performing a long-running task, it should give feedback to the user of how far the job has progressed, and provide a means to cancel it.

Typically, `JProgressMonitor` is used for this. `JProgressMonitor` is not a Swing component itself, but shows a dialog with a progress bar in it when the task has taken more than half a second (this is configurable). The following figure shows the user interface of a `JProgressMonitor`:

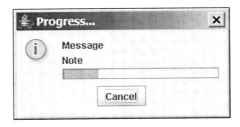

By default, the dialog contains a `JProgressBar`, two labels and a **Cancel** button.

More Techniques for Testing Swing Components

If our software is expected to show a progress monitor, we will want to check:

- Is the monitor actually showing?
- Are the message and note strings showing as expected?
- Is the amount of progress as expected?
- What happens if we cancel the operation?

To help write this kind of test, we have written a class, UIProgressMonitor. Not only is this a useful class, it is also a good example of how, with sufficient determination, we can find out a lot about the state of an object (in this case a ProgressMonitor) without actually having direct access to it.

The way UIProgressMonitor works is as follows. Assuming that there is only one JProgressBar showing, we can find it using our standard UI search tricks. Having the JProgressBar, we can read the maximum, minimum, and current values.
To get the note and message strings, we can climb the containment tree from the JProgressBar. This should yield a JPanel, which in turn contains a JLabel each for the message and the note, from which the string values can be read:

```java
public class UIProgressMonitor {
    private JProgressBar bar;
    private JLabel messageLabel;
    private JLabel noteLabel;
    public UIProgressMonitor() {
        final JProgressBar[] barHolder = new JProgressBar[1];
        Waiting.waitFor( new Waiting.ItHappened() {
            public boolean itHappened() {
                barHolder[0] =
                        UI.findProgressBarThatIsCurrentlyShowing();
                return barHolder[0] != null;
            }
        }, 10000 );
        bar = barHolder[0];
        assert (bar != null) : "Could not find a bar!";
        //From the JProgressBar we can climb the containment
        //tree to the parent, which directly contains the
        //labels in which we are interested.
        Container parent = bar.getParent();
        Component[] allComponents = parent.getComponents();
        messageLabel = (JLabel) allComponents[0];
        noteLabel = (JLabel) allComponents[1];
    }
    ...
}
```

Chapter 10

This code is fragile, in that it is assuming a lot of internal details about the `ProgressMonitor` class. However, this should not stop us from using `UIProgressMonitor` in tests as we can always fix the code if things change.

We might typically use `UIProgressMonitor` like this:

```
//Do some action that initiates a lengthy task.
//...etc.etc.

//Create a UIProgressMonitor and check details.
UIProgressMonitor uipg = new UIProgressMonitor();
assert uipg.message().equals( expectedMsg );
assert uipg.getNot().equals( expectedNote );
assert uipg.getMinimum() == 0;
assert uipg.getMaximum() == 100;
...etc.

//Wait for a certain amount of progress and then cancel.
uipg.waitForProgress( 50 );
uipg.cancel();

//Check that the task has actually been cancelled.
//...etc.
```

The unit tests for `UIProgressMonitor` give more examples of its use.

JSlider and JSpinner

'Ikon Do It' provides a tool for setting the transparency of the pixel currently being edited either with a `JSpinner` or a `JSlider`:

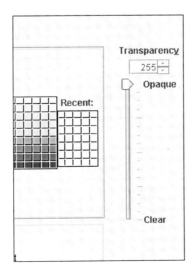

The UI wrapper for this control is `UITransparencySelecter`:

```
public class UITransparencySelecter {
    Cyborg robot = new Cyborg();
    private IkonMakerUserStrings us =
                IkonMakerUserStrings.instance();
    public void setValue( int value ) {
     robot.altChar( us.mnemonic(
            IkonMakerUserStrings.TRANSPARENCY ) );
        robot.selectAllText();
        robot.type( "" + value );
        robot.enter();
    }
    /**
     * Checks that the spinner and slider show
     * the same value and returns it as an int.
     */
    public int value() {
        int spinnerValue = Integer.parseInt(
                UI.getSpinnerValue( spinner() ).toString());
        int sliderValue = UI.getSliderValue( slider() );
        assert sliderValue == spinnerValue;
        return spinnerValue;
    }
    public JSlider slider() {
        return (JSlider) UI.findNamedComponent(
                    TransparencySelecter.SLIDER_NAME );
    }
    public JSpinner spinner() {
        return (JSpinner) UI.findNamedComponent(
                    TransparencySelecter.SPINNER_NAME );
    }
}
```

The `setValue(int)` method shows how to set a spinner value using `Cyborg`, while `value()` allows the test to safely read spinner and slider values. Both values are read so that a check that the spinner and slider agree can be made.

JTree

If a `JTree` has the focus, then it can be navigated with the *Home, down arrow,* and *right arrow* keys. The *Home* key selects the top node. If the root node is showing, this will be the root, otherwise it will be the first child of the root. The right arrow key expands the currently selected node if it is collapsed, moves to the first child of the current node if there is one, and does nothing on leaf nodes.

For example, the tree in the following figure is a JTree with the root node selected:

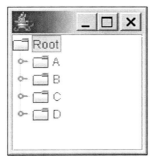

Pressing the combination 'right arrow, right arrow, down arrow' 10 times gives:

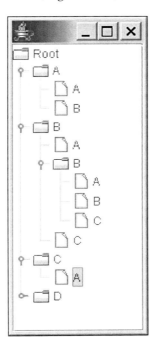

This fact makes it easy to write a Cyborg method to select a JTree node based on its position in a depth-first traversal of the tree:

```
//From Cyborg.
/**
 * Expands at least as many tree nodes are required,
 * and then arrows down to the indexed item.
 */
```

```
public void selectTreeItemByDepthFirstIndex( int index ) {
    selectTreeRoot();
    int maxNeeded = 2 * index;
    for (int i=0;i < maxNeeded; i++){
        right();
        right();
        down();
    }
    selectTreeRoot();
    for (int i=0;i < index; i++){
        down();
    }
}
/**
 * Selects the root node of the currently
 * focussed tree.
 */
public void selectTreeRoot() {
    home();
}
```

This code first selects the top node and then does enough right-arrow and down-arrow keystrokes to ensure that sufficient tree nodes are expanded to make the required node visible. At this point, we don't know which node is selected, so we go back to the top of the tree and use the down-arrow to the node we want.

Complementary to this `Cyborg` method for selecting a tree node is a `UI` method for safely reading the selected path for a tree — `UI.getSelectedTreePath(JTree)`.

Summary

This chapter completes our discussion of unit testing of user interface components. The `Cyborg` and `UI` classes provide a rich library for writing test code, and the unit tests for these classes themselves provide many examples of safely testing user interfaces. Above all, these examples should demonstrate that it is possible to manipulate and test almost any component in a thread-safe manner, even without programmatic access to the component itself.

11
Help!

As more and more features have been added to LabWizard, the **Help** system for it has become extensive and now contains about three hundred HTML files plus a similar number of images. Even before we had so much data to manage, we were faced with these problems:

- There were broken links in the help index.
- Some files were not listed in the index, so they could not be read by customers.
- Some of the contextual help buttons showed the wrong topic.
- Some of the HTML files contained broken links or incorrect image tags.
- Not all of the file titles matched their index entry.

The last problem manifested itself when the user did a free text search of the help system. The result of such a search is a list of files, each represented by its title. With our system, most of the documents had the title **untitled**.

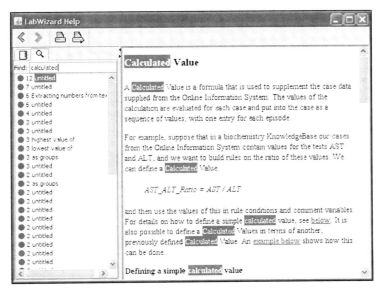

In fact, the JavaHelp 2.0 System User's Guide contains the recommendation

> *"To avoid confusion, ensure that the* `<TITLE>` *tag corresponds to the title used in the table of contents."*

Given that customers mostly use the Help system when they are already frustrated by our software, we decided that it was very important to fix these errors. To do this, we wrote a tool, `HelpGenerator`, that generates some of the boilerplate XML in the help system and checks the HTML and index files for the problems listed above. We also built tools for displaying and testing the contextual help. We've re-engineered and improved these tools and present them here.

In this chapter we are assuming familiarity with the JavaHelp system. Documentation and sample code for JavaHelp can be found at: http://java.sun.com/products/javahelp.

Overview

A JavaHelp package consists of:

- A collection of HTML and image files containing the specific Help information to be displayed.
- A file defining the index of the Help topics. Each index item in the file consists of the text of the index entry and a string representing the target of the HTML file to be displayed for that index entry, for example:

    ```
    <index version="1.0">
      <indexitem text="This is an example topic."
                 target="Topic">
        <indexitem text="This is an sub-topic."
                   target="SubTopic"/>
      </indexitem>
    </index>
    ```

- A file associating each target with its corresponding HTML file (or more generally, a URL) — the map file. Each map entry consists of the target name and the URL it is mapped to, for example:

    ```
    <map version="1.0">
      <mapID target="Topic" url="Topic.html"/>
      <mapID target="SubTopic" url="SubTopic.html"/>
    </map>
    ```

- A HelpSet file (by default `HelpSet.hs`) which specifies the names of the index and map files and the folder containing the search database.

Our software will normally have a main menu item to activate the Help and, in addition, buttons or menu items on specific dialogs to activate a Help page for a particular topic, that is, "context-sensitive" Help.

What Tests Do We Need?

At an overall structural level, we need to check:

- For each target referred to in the index file, is there a corresponding entry in the map file? In the previous example, the index file refers to targets called `Topic` and `SubTopic`. Are there entries for these targets in the map file?
- For each URL referred to in the map file, is that URL reachable? In the example above, do the files `Topic.html` and `SubTopic.html` exist?
- Are there HTML files in our help package which are never referred to?
- If a **Help** button or menu item on some dialog or window is activated, does the Help facility show the expected topic?
- If the Help search facility has been activated, do the expected search results show? That is, has the search database been built on the latest versions of our Help pages?

At a lower level, we need to check the contents of each of the HTML files:

- Do the image tags in the files really point to images in our help system?
- Are there any broken links?

Finally, we need to check that the contents of the files and the indexes are consistent:

- Does the title of each help page match its index?

To simplify these tests, we will follow a simple naming pattern as follows:

We adopt the convention that the name of each HTML file should be in **CamelCase** format (conventional Java class name format) plus the `.html` extension. Also, we use this name, without the extension, as the target name. For example, the target named `SubTopic` will correspond to the file `SubTopic.html`.

Furthermore, we assume that there is a single Java package containing all the required help files, namely, the HTML files, the image files, the index file, and the map file. Finally, we assume a fixed location for the Help search database.

With this convention, we can now write a program that:

- Generates the list of available targets from the names of the HTML files.
- Checks that this list is consistent with the targets referred to in the index file.
- Checks that the index file is well-formed in that:
    - It is a valid XML document.
    - It has no blank index entries.
    - It has no duplicate index entries.
    - Each index entry refers to a unique target.
- Generates the map file, thereby guaranteeing that it will be consistent with the index file and the HTML files.

The class `HelpGenerator` in the package `jet.testtools.help` does all this, and, if there are no inconsistencies found, it generates the map file. If an inconsistency or other error is found, an assertion will be raised. `HelpGenerator` also performs the consistency checks at the level of individual HTML files. Let's look at some examples.

An HTML File That is Not Indexed

Here is a simple help system with just three HTML files:

The index file, **HelpIndex.xml**, only lists two of the HTML files:

```
<index version="1.0">
  <indexitem text="This is an example topic."
             target="ExampleTopic">
    <indexitem text="This is an example sub-topic."
               target="ExampleSubTopic"/>
  </indexitem>
</index>
```

When we run `HelpGenerator` over this system (we'll see how to do this later in this chapter), we get an assertion with the error message **The Help file: TopicWithoutTarget.html was not referenced in the Index file: HelpIndex.xml**.

An index item for which there is no HTML file

If we take the help system in the previous example, but remove the file **TopicWithoutTarget.html** (which was not indexed) and the file **ExampleSubTopic.html** (which did have an index item) then we are left with a system in which the index file references a non-existent HTML file. Applying `HelpGenerator` to this system gives an assertion with the error message **The entry: ExampleSubTopic in the Index file:HelpIndex.xml did not have a HTML file**.

Broken links

Consider this simple help system. The following screenshot shows a help system in which some of the links in the HTML files are broken.

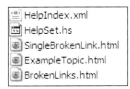

Running `HelpGenerator` over this system produces an error that indicates which files have broken links and which links are actually broken:
The file BrokenLinks.html has these broken links:
NonExistent1.html
NonExistentImage.png
The file SingleBrokenLink.html has this broken link:
NonExistent1.html.

Incorrectly titled help pages

As a final example, consider a help system consisting of three HTML files `A.html`, `B.html`, and `C.html` such that

- `A.html` has an empty title tag.
- `B.html` has title "Topic B".
- `C.html` has title "untitled".

Further, suppose that our index file is as follows:

```
<index version="1.0">
  <indexitem text="Topic A" target="A"/>
  <indexitem text="Topic B" target="B"/>
  <indexitem text="Topic C" target="C"/>
</index>
```

The `HelpGenerator` tool applied to this system fails with an error message that indicates the problem, which is:
"Title" does not match index text 'Topic A' for file A.html
Title 'untitled' does not match index text 'Topic C' for file C.html.

Creating and Testing Context-Sensitive Help

Each component of an application should include a facility for showing the Help system with the topic most relevant to that component selected. For example, the LabWizard Attribute Manager screen has a context-sensitive **Help** button:

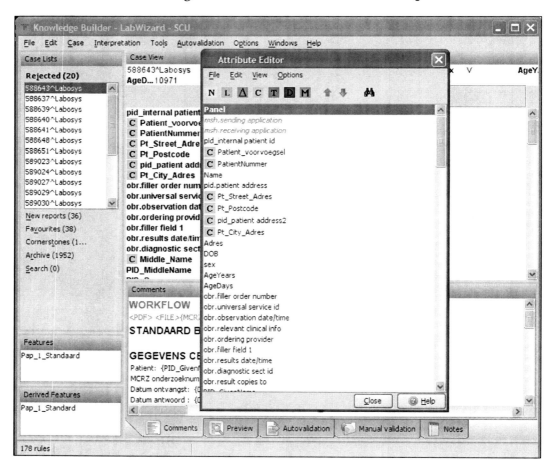

Activating this button shows the **Attributes** chapter of the Help system:

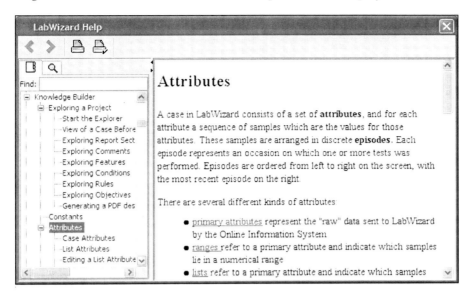

`HelpGenerator` generates an enumeration, `Help.java`, which makes it very easy to include this kind of context-sensitive help in our application screens. For example, the code to build the **Help** button in the **Attribute Editor** shown previously is:

```
private Box createButtonBar() {
    JButton closeButton = new JButton( actions.close );
    Box box = Box.createHorizontalBox();
    box.add( Box.createHorizontalGlue() );
    box.add( closeButton );
    box.add( Box.createHorizontalStrut( 5 ) );
    box.add( createHelpButton( Help.Attribute ) );
    return box;
}
```

The highlighted line here creates the button and adds it to the button box at the bottom of the dialog. The code to create the button itself is:

```
public static JButton createHelpButton( final Help topic ) {
    JButton helpButton = us.createButton( SimpleUIMessages.HELP );
    helpButton.addActionListener( new ActionListener() {
        public void actionPerformed( ActionEvent e ) {
            HelpGenerator.helpActionFor(
                            topic ).actionPerformed( e );
        }
    } );
    return helpButton;
}
```

Help!

The first line of this method uses the `UserStrings` class introduced in Chapter 5 to create the button, and the second line adds an `ActionListener` to the button that actually shows the Help browser with correct topic shown. Here is the `HelpGenerator` method that creates the `ActionListener`:

```
public static ActionListener helpActionFor( Help topic ) {
    if (broker == null) {//broker is a javax.help.HelpBroker
        initialiseBroker();
    }
    broker.setCurrentID( topic.toString() );
    return new CSH.DisplayHelpFromSource( broker );
}
```

The enumeration `Help.java` lists all of the available help topics. For example, a Help system containing just the HTML files `ExampleTopic.html` and `ExampleSubTopic.html` would be represented as:

```
package book.testtools.help;

public enum Help {
    ExampleSubtopic,
    ExampleTopic,
}
```

This enumeration is automatically generated by `HelpGenerator.main()` method, provided that all the consistency checks described earlier pass.

Using this methodology we can be sure that activating a **Help** button or menu item will bring up a Help topic. However, without inspecting the source code, how can we check whether the correct topic has been associated with this **Help** button or menu item?

Our unit tests can be based on the `HelpGeneratorTest` method:

```
public static void checkHelpIsShowingForTopic( Help topic )
```

This method reads the index file to determine the index item for the specified topic, finds the frame showing the help facility's index `JTree`, and then checks whether this index item is in fact selected. If so, then we can be sure from the other `HelpGenerator` checks that the appropriate HTML file is displayed. If the required index item is not selected then an assertion is raised.

As an example of how easy these unit tests are to write, here is the test for the **Help** button in the **Attribute Editor** unit test:

```
public boolean helpTest() {
    init();
    robot.help();
    HelpGeneratorTest.checkHelpIsShowingForTopic(
                              Help.Attribute );
    cleanup();
    return true;
}
```

In this test, the first line calls the `init()` method that populates and displays an Attribute Editor. The second line uses a mnemonic to activate the **Help** button. The third line uses the `HelpGeneratorTest` method to check that the JavaHelp browser is showing and is displaying the correct topic.

More details of `HelpGenerator` and `HelpGeneratorTest` can be found in the packages `jet.testtools.help` and `jet.testtools.help.test`.

Executing HelpGenerator

To execute `HelpGenerator`, two parameters need to be provided: the name of the package we are using to provide the help functionality, and the name of the corresponding source directory, for example:

```
private String helpPackageName = "jet.run.help";
private String helpDirectoryName = "C:\\jet\\run\\help";
HelpGenerator.main( new String[]{helpPackageName,
                        helpDirectoryName});
```

As `HelpGenerator` creates a production class, it should be executed with each software build. Here is our Ant task for doing this. Note that we rebuild the search database at the same time so that it is always up to date:

```
<target name="GenerateHelp" description="Generate Help info.">
    <java classname="jet.testtools.help.HelpGenerator" fork="true"
        maxmemory="384m" dir="${src}">
        <classpath>
            <path refid="classpath" />
        </classpath>
        <arg value="jet.testtools.help"/>
        <arg path="${src}/jet/testtools/help"/>
    </java>
```

```
        <java classname="com.sun.java.help.search.Indexer" fork="true"
    maxmemory="384m" >
            <arg line="${helpsrc}/*.html"/>
            <arg line="-db ${helpdst}/JavaHelpSearch"/>
            <arg line="-c ${src}/IndexerConfig.txt"/>
            <classpath>
                <path refid="classpath" />
            </classpath>
        </java>
    </target>
```

In addition, `HelpGenerator` can be executed whenever a new Help page is added, enabling the corresponding topic to be used by a **Help** button or menu item.

Summary

By using `HelpGenerator` and its associated test class, we can be sure that our help system has no broken indexes, has no missed topics, and shows the right topic for each context. As well, we will know that our help files contain no broken links, no missing images, and are correctly titled. We can also use `HelpGenerator` and `HelpGeneratorTest` to greatly simplify the implementation and testing of our application's context-sensitive help.

12
Threads

With more demanding customer needs and the greater availability of multi-core systems, server applications have become increasingly multi-threaded over the years. For example, in our LabWizard server there are threads for:

- Receiving cases from the lab system.
- Sending reports to the lab system.
- Periodic tasks such as committing the databases and doing a nightly backup.
- Logging.
- Sending emails.
- Communicating with users via RMI.

Additionally, some of the client commands create their own threads on the server in which data mining tasks are performed.

Although we've never had quite as much time as we'd like to carefully think through the concurrent programming involved, the software has very few concurrency problems and no serious ones have been reported yet. What we have managed to do is write some very rigorous tests that exercise the system in as many ways as possible at once, and fix any problems that these tests have uncovered.

This chapter explains some of the tools and techniques that we have used in our testing of the concurrent behavior of LabWizard. The methods we introduce here are all simple, but they are very effective and can be used in a wide variety of situations.

The Waiting Class

We often need our test thread to suspend execution for some length of time. If we know how long this should be, we can just call `Thread.sleep(long time)`.

More commonly, though, we need a test to wait for some event, such as an email being received, or a socket connection being made, and we don't know how long to wait for this to happen. In this situation, we could work out a reasonable upper limit for the time taken, and just wait for that period. This approach would be problematic because we would either have very flaky tests or very slow tests or, more likely, tests that are both slow and flaky.

If we are not conservative in guessing these wait periods, we are liable to have tests that fail occasionally. This is because sometimes the time the test waits turns out not to be quite long enough. Such failures can be caused by the tests being run on a slower machine than that on which they were first written, because of suspension of the test thread while the JVM does garbage collection, or for a variety of other reasons.

On the other hand, if we are very conservative in our waiting periods, we will slow our tests down unacceptably. For example, if we wait ten seconds for a socket connection that might actually happen within two seconds, we are throwing away eight seconds of testing time. This wasted time will not be reduced when we move our tests to faster machines, unless we laboriously adjust our guesses in all tests with such pauses. With a large system, where we are already struggling to run the tests in a reasonable amount of time, we cannot afford to slow things down with unnecessary pauses.

This trade-off between robustness and test speed is the wrong game to be playing. A better solution is to use the method

```
Waiting.waitFor( Waiting.ItHappened ih, long timeout )
```

which waits for a specified length of time for an event to occur. If the event happens, the method returns `true`. If the wait times out, the result is `false`. The interface `Waiting.ItHappened` is used to express the condition on which our test thread is to wait.

Suppose, for example, that our test needs to wait for a directory to be empty. We can do this easily:

```
final File outputDir = ...;
boolean wasCleared = Waiting.waitFor( new Waiting.ItHappened() {
    public boolean itHappened() {
        return outputDir.listFiles().length == 0;
    }
}, 10000 );
```

This will wait for up to 10 seconds for outputDir to be empty. If the directory is cleared within 10 seconds, wasCleared will be true, otherwise false. We can make assertions about the return value, if we need to. The waitFor method suspends the calling thread and starts a new thread in which the checks occur. The new thread checks the ItHappened condition every 10 milliseconds.

Sometimes, the check we want to make uses a lot of resources, or takes a lot of time, so we don't want to check too frequently (for "a watched pot never boils", as TL's father says). In this situation, we can set the checkPeriod between checks using:

```
Waiting.waitForCheckingWithPeriod(
        Waiting.ItHappened ih, long timeout, long checkPeriod );
```

The Waiting class we're using here is in the package, jet.testtools. In addition to the methods we've already seen, it declares the pause(long millis) method, which is a handy, exception-suppressing wrapper for Thread.sleep(long millis). Another useful method is Waiting.stop(), which suspends the calling thread indefinitely, and can be used to stop a broken test so that the state of the system can be inspected.

Concurrent Modifiers

Suppose there is an object in our production system that may be modified concurrently by multiple threads. In the unit test for the corresponding class, we need to check whether it is in fact "thread-safe". The basic approach is to construct a test harness with a set of threads concurrently modifying the state of the object as intensively as possible, checking whether the behavior of the object remains consistent in this environment.

For example, suppose we have a class IntegerFactory whose task it is to generate a unique integer whenever next() is called, as shown below:

```
public class IntegerFactory {
    private static Integer current = -1;
    public static Integer next() {
        return ++current;
    }
    //Private constructor as we don't want this instantiated.
    private IntegerFactory() {
    }
}
```

Threads

To test this method, we create accessor threads (instances of the `Reader` class below) whose task it is to call the `next()` method a specified number of times, say N. We therefore expect each accessor thread to read N unique integers. Further, no integer should be read by more than one `Reader`, there should be no gaps in the integers read by all the readers, and the integers read by any reader should be in ascending order. Once the threads have all completed we can make these checks:

```java
public class IntegerFactoryTest {
    final int NUMBER_OF_READERS = 3;
    final int NUMBER_TO_READ = 10000;

    public boolean nextTest() throws Exception {
        Reader[] readers = new Reader[NUMBER_OF_READERS];
        for (int i = 0; i < readers.length; i++) {
            readers[i] = new Reader();
            readers[i].start();
        }
        for (Reader reader : readers) {
            reader.join();
        }
        //Collect together all the values that were read.
        SortedSet<Integer> allValuesRead = new TreeSet<Integer>();
        //Record and check the values for each reader.
        for (Reader reader : readers) {
            for (int i = 0; i < NUMBER_TO_READ; i++) {
                //Check that the value
                //was not read by another reader.
                assert !allValuesRead.contains( reader.values[i] );
                //Record the value as read.
                allValuesRead.add( reader.values[i] );
                //Check that the values are increasing.
                if (i > 0) {
                    assert reader.values[i] > reader.values[i - 1];
                }
            }
        }
        assert allValuesRead.first() == 0;
        int expectedCount = NUMBER_OF_READERS * NUMBER_TO_READ;
        assert allValuesRead.size() == expectedCount;
        assert allValuesRead.last() == expectedCount - 1;

        return true;
    }

    private class Reader extends Thread {
        int[] values = new int[NUMBER_TO_READ];
```

```
        public void run() {
            for (int i = 0; i < NUMBER_TO_READ; i++) {
                values[i] = IntegerFactory.next();
            }
        }
    }
}
```

With just one instance of the `Reader`, the test of course never finds any problems. With NUMBER_OF_READERS = 3 and NUMBER_TO_READ = 100, the test occasionally triggers an assertion. When NUMBER_TO_READ is sufficiently large, say 10000, the test always throws an assertion error, thus providing a good basis for testing any fix that we might implement (such as making the method synchronized).

Concurrent Readers and Writers

A more complex concurrent modifiers situation is when a class provides separate read and write operations. The testing approach is similar to before, but now we need to check that both readers and writers can safely access the class concurrently.

For example, suppose we have a `Pool` class which provides a `put()` method to store an object and a `getAll()` method to retrieve all objects currently stored. In this example we implement `Pool` as a simple wrapper around a `Set`:

```
public class Pool<E> {
    private Set<E> set = new HashSet<E>();

    public synchronized void put( E e ) {
        set.add( e );
    }
    public synchronized Set<E> getAll() {
        Set<E> result = new HashSet<E>();
        result.addAll( set );
        return result;
    }
}
```

To test the `Pool` class, we create some `Reader` threads whose task is to call the `getAll()` method N times and some `Writer` threads whose task is to call the `put()` method also N times, each time storing some arbitrary data in the `Pool`:

```
public class PoolTest {
    final int INSTANCES = 10;
    final int ITERATIONS = 100;
    private Pool<String> pool;
```

```
    ...
    private class Reader extends Thread {
        boolean finished = false;
        public Reader( int i ) {
            super( "Reader" + i );
        }
        public void run() {
            for (int i = 0; i < ITERATIONS; i++) {
                pool.getAll();
            }
            finished = true;
        }
    }
    private class Writer extends Thread {
        boolean finished = false;
        public Writer( int i ) {
            super( "Writer " + i );
        }
        public void run() {
            for (int i = 0; i < ITERATIONS; i++) {
                pool.put( getName() + " Count:   " + i );
            }
            finished = true;
        }
    }
}
```

Once each thread has completed, our main test thread simply checks whether each Reader and Writer thread managed to do its task without generating an exception:

```
public class PoolTest {
    ...
    public boolean concurrencyTest () throws Exception {
        pool = new Pool<String>();
        Reader[] readers = new Reader[INSTANCES];
        Writer[] writers = new Writer[INSTANCES];
        for (int i = 0; i < INSTANCES; i++) {
            readers[i] = new Reader( i );
            writers[i] = new Writer( i );
            readers[i].start();
            writers[i].start();
        }
```

```
            for (Reader reader : readers) {
                reader.join();
                assert reader.finished:
                        reader.getName() + " did not finish";
            }
            for (Writer writer : writers) {
                writer.join();
                assert writer.finished:
                        writer.getName() + " did not finish";
            }
            return true;
        }
        ...
    }
```

If any of the test threads had generated an exception, they would have terminated without setting their finished flag to true.

This test reliably demonstrates the problems with Pool, with the test threads generating ConcurrentModificationErrors every time the test is run. With such a reliable test, we can be reasonably confident that when we change the class so that it passes the test, we really have fixed the concurrency problem.

Proof of Thread Completion

In the LabWizard Knowledge Builder, it is possible to search amongst the cases for those that satisfy some condition. Often, the user wants to cancel this operation, either because they've realized that they're using the wrong search condition, or because enough cases have already been found. On the server side, each search operation is carried out in its own thread. If the user cancels the operation, this thread is stopped. (Actually, a variable is set that tells the thread that it should stop searching and complete its run() method. The stop() method of Thread is deprecated.)

In our tests for this, we need to check that if a search operation is cancelled, the thread in which it is running terminates promptly. How can we write such a test?

It is easy to check that a thread completes if we give it a name. Before we start an operation, we check that no thread with that name exists within the JVM. Once the operation is started, we check that the named thread is active. We then cancel the operation and check that the thread is removed from the list of active threads within some reasonable length of time.

The list of active threads in a JVM can be found using the methods `Thread.activeCount()` and `Thread.enumerate()`. We've provided a useful extension for these in our `TestHelper` class:

```
//From TestHelper.
public static Set<String> namesOfActiveThreads() {
    Set<String> result = new HashSet<String>();
    Thread[] threads = new Thread[Thread.activeCount()];
    Thread.enumerate( threads );
    for (Thread thread : threads) {
        if (thread != null) {
            result.add( thread.getName() );
        }
    }
    return result;
}
```

By combining this with our methods for waiting, we get this useful method:

```
//From TestHelper.
public static boolean waitForNamedThreadToFinish(
        final String threadName, long timeoutMillis ) {
    Waiting.ItHappened ih = new Waiting.ItHappened() {
        public boolean itHappened() {
            return !namesOfActiveThreads().contains( threadName );
        }
    };
    return Waiting.waitFor( ih, timeoutMillis );
}
```

Let's see these methods put to use in a real test. Suppose we have a class `Server` that has a `housekeeping()` method which zips all the files in an output directory. We don't want this task to interfere with normal server processing, so the housekeeping should be done in its own thread:

```
public class Server {
    public static final String HOUSEKEEPING = "Housekeeping";
    private File inputDir;
    private File outputDir;
    private File archiveDir;
    public Server( File inputDir,
                    File outputDir, File archiveDir ) {
        this.inputDir = inputDir;
        this.outputDir = outputDir;
        this.archiveDir = archiveDir;
    }
    public void housekeeping() {
```

```
            new Thread( HOUSEKEEPING ) {
                public void run() {
                    File[] filesToZip = outputDir.listFiles();
                    File zipFile = new File( archiveDir,
                                             "archived.zip" );
                    Zipper.zip( filesToZip, zipFile );
                    //delete the original files
                    for (File file : filesToZip) {
                        file.delete();
                    }
                }
            }.start();
        }
        ...
    }
```

Our test for the housekeeping method needs to do the following:

- Put some known test files into the output directory.
- Call the `housekeeping()` method.
- Wait for the corresponding thread to complete.
- Check that the zip file has been created, that it contains the expected test files, and that the output directory is now empty. Here is the test:

```
public class ServerTest {
        private File inputDir;
        private File outputDir;
        private File archiveDir;
        private Server server;
        private void init() {
            //Clean up any files from a previous test.
            File tempDir = Files.cleanedTempDir();
            outputDir = new File( tempDir, "outputDir" );
            outputDir.mkdirs();
            inputDir = new File( tempDir, "inputDir" );
            inputDir.mkdirs();
            archiveDir = new File( tempDir, "archiveDir" );
            archiveDir.mkdirs();
            server = new Server( inputDir, outputDir, archiveDir );
        }
        public boolean housekeepingTest() throws Exception {
            init();
            //Put 2 configured files into the output directory.
            File file1 = new File( Testtools.datafile1_txt );
            File file2 = new File( Testtools.datafile2_txt );
```

```
            Files.copyTo( file1, outputDir );
            Files.copyTo( file2, outputDir );
            //Sanity check that these are the only 2 files in it.
            Assert.equal( outputDir.listFiles().length, 2 );
            //run the housekeeping task
            server.housekeeping();

            //wait for it to complete
            boolean completed = TestHelper.waitForNamedThreadToFinish(
                    Server.HOUSEKEEPING, 2000 );
            assert completed : "Housekeeping thread did not stop!";
            //check that the archive file has been created
            File expectedZip = new File( archiveDir, "archived.zip" );
            assert expectedZip.exists();

            //Check that the 2 files
            //put into the output dir have been deleted.
            Assert.equal( outputDir.listFiles().length, 0 );
            //Check the contents of the zip files.
            Files.unzip( expectedZip.getPath(), Files.tempDir() );
            File retrieved1 = new File( Files.tempDir(),
                                        file1.getName() );
            File retrieved2 = new File( Files.tempDir(),
                                        file2.getName() );
            Assert.equal( Files.contents( file1 ),
                    Files.contents( retrieved1 ) );
            Assert.equal( Files.contents( file2 ),
                    Files.contents( retrieved2 ) );
            cleanup();
            return true;
        }
        //Other tests follow.
    }
```

Apart from allowing us to test thread completion, having names for the threads we create is also really useful in the investigation of tests that do not terminate (as discussed in Chapter 20). For these two reasons, we recommend that it always be done:

Extreme Testing Guideline: Each thread our software creates should be given a name.

The Unit Test for waitForNamedThreadToFinish()

The test for `waitForNamedThreadToFinish()` is itself a good example of how to work with threads in tests. To test this method, we need a way of creating named threads that can be readily forced to finish. This is achieved by a simple extension of `Thread`:

```
public class TestHelperTest {
    ...
    private class NamedThread extends Thread {
        boolean flagToStop;
        public NamedThread( String name ) {
            super( name );
        }
        public void run() {
            while(!flagToStop) {
                Waiting.pause( 10 );
            }
        }
    }
}
```

Our test will create `NamedThread`s and call `TestHelper.waitForNamedThreadToFinish()`. In order for the wait to succeed, rather than time out, we need `flagToStop` to be set during the wait itself. One way of doing this is to have a thread for calling `waitForNamedThreadToFinish()`:

```
public class TestHelperTest {
    ...
    private class ThreadInWhichToCallWaitFor extends Thread {
        boolean result;
        String threadName;
        long timeout;
        public ThreadInWhichToCallWaitFor(
                String threadName, long timeout ) {
            super( "WaiterFor: " + threadName );
            this.threadName = threadName;
            this.timeout = timeout;
        }
        public void run() {
            result = TestHelper.waitForNamedThreadToFinish(
                    threadName, timeout );
        }
    }
}
```

The test is now simplicity itself:

```
public class TestHelperTest {
    ...
    public boolean waitForNamedThreadToFinishTest()
                                    throws Exception {
        //One where the thread finishes within the expected period.
        NamedThread aubrey = new NamedThread( "Aubrey" );
        ThreadInWhichToCallWaitFor waiter
                = new ThreadInWhichToCallWaitFor( "Aubrey", 1000 );
        aubrey.start();
        waiter.start();
        Waiting.pause( 100 );
        aubrey.flagToStop = true;
        waiter.join();
        assert waiter.result;

        //One where the thread does not finish
        //within the expected period.
        NamedThread maturin = new NamedThread( "Maturin" );
        waiter = new ThreadInWhichToCallWaitFor( "Maturin", 1000 );
        maturin.start();
        waiter.start();
        waiter.join();
        assert !waiter.result;
        //Try again. This is partly just so that maturin stops.
        waiter = new ThreadInWhichToCallWaitFor( "Maturin", 1000 );
        waiter.start();
        maturin.flagToStop = true;
        waiter.join();
        assert waiter.result;
        return true;
    }
}
```

Note the semi-redundant check in the last part of the test. If we don't make the call

```
maturin.flagToStop = true;
```

the JVM does not terminate after the test has finished. We could make the threads 'daemon' threads, but there is some value in checking that a successful wait can follow a failed one.

Counting Threads

Sometimes we want to see how many threads are running in our JVM. As an example, suppose our software has a method which uses several concurrently executing threads. We need to check whether the required number of threads are actually running, and that they have all terminated once the task has been performed.

We can use the method `Thread.activeCount()` to do these checks.

For example, suppose our `Server` has an `archive()` task which creates five concurrently executing threads. Each thread terminates itself once the task has been completed.

Our test for the `archive()` method needs to do the following:

- At the start of the test, count the number of active threads. This will be more than one if some user interface elements have been shown in previous tests.
- Call the `archive()` method.
- Check that the five additional threads have been created.
- Wait for the task to complete. (In this example, we expect the task to take no more than 10 seconds on our test platform.)
- Check that the additional threads have terminated.
- Check whether the archiving task has indeed been performed as expected.

```
public class ServerTest {
    public boolean archiveTest() {
        init();
        //Get an initial count of the active threads.
        final int numThreadAtStart = Thread.activeCount();

        //Start the archiving process. This takes
        //at least 5 seconds.
        server.archive();

    //Check whether we now have 5 additional threads.
    //There is really no race condition here as server.archive()
    //called above takes at least 5 seconds.
    Assert.equal( Thread.activeCount(), numThreadAtStart + 5 );

        //Wait for 10 secs for all archive threads to complete
        Waiting.waitFor( new Waiting.ItHappened() {
            public boolean itHappened() {
                return Thread.activeCount() == numThreadAtStart;
            }
        }, 10000 );
        //Check whether the archiving was performed....
        return true;
    }
}
```

Summary

In this chapter, we have looked at several techniques for testing multi-threaded systems. First off, we looked at the Waiting class, which can be used to suspend the testing thread until some event occurs. Next, we looked at a simple but effective technique for testing concurrent readers and writers. Finally, we looked at tests of the actual threads in the JVM—testing that a known thread has finished and testing the number of active threads.

Further Reading

It's worth having a look at the class jet.testtools.Waiting and its unit test. The Server and Pool classes are in jet.concurrencyexamples and are also worth a look, as are their unit tests.

 The book *Java Concurrency in Practice* (Goetz et al, Addison Wesley, 2006) contains useful chapters on unit and performance tests for multi-threaded applications.

13
Logging

Most server programs will perform some kind of logging to record their interactions with users and other systems. With logging frameworks such as the one built into the JDK, logging is so convenient and efficient that it is easy for the use of log files to become an essential part of the deployed software, used in billing or user tracking, for example. This 'scope-creep' is alright as long as it is documented and above all, tested.

In this chapter we will look at techniques for testing what is logged by our software. We will also show that it is possible to check what is being printed to the console by our programs.

The examples that follow are built around the logging package, `java.util.logging`, built into the JDK, but would work with any of the other popular logging packages, such as Log4J.

Logging to a File

Suppose our system is required to log various transactions and any error messages. To check whether this is happening, our test will generate a transaction or error condition and then wait for the logger to record the expected entry.

As an example, suppose our system uses a logger with a file handler. Our system will use the one provided by the class `jet.testtools.FileLogger`, which is a simple wrapper for the Java logging package. A simple test would be:

```
    public boolean logTest() {
        //Roll the log to clear it.
        FileLogger.rollLogOver();
        //Call the method that should generate a log entry.
        final String entry = "This is a normal transaction";
        productionMethodThatGeneratesAnEntry( entry );
```

```
        //Wait for the entry to appear in the log file.
        final File logFile =
                new File( FileLogger.logDirName, "log.txt.0" );
        Waiting.waitFor( new Waiting.ItHappened() {
            public boolean itHappened() {
                return Files.contents( logFile ).contains( entry );
            }
        }, 1000 );
        return true;
    }
```

Note that we have expected the entry to be written to the log file within one second.

Suppose we have the slightly more complex situation where our production system had a bug which would generate an exception, which in turn would be logged. The test code for our bug fix will need to check that the exception is not generated, as shown by the entry no longer appearing in the log file. How long should we wait for the entry before we are satisfied that it is not going to appear?

A simple way to approach this issue is to log a marker entry after the exception would have been generated, and then to wait for the marker entry to appear. We then search the log file for the exception entry. If the exception has not been logged by this time, we can be sure it's not going to be. For example:

```
    public boolean exceptionNoLongGeneratedTest() {
        //Roll the log to clear it.
        FileLogger.rollLogOver();
        //Calling this method used to cause an exception
        //stack trace to be written in the log file.
        productionMethodThatUsedToGenerateException( false );

        //Log a marker entry.
        final String marker = "This is a marker entry";
        FileLogger.log( new LogRecord( Level.INFO, marker ) );
        //Wait for the marker to appear.
        final File logFile =
                new File( FileLogger.logDirName, "log.txt.0" );
        Waiting.waitFor( new Waiting.ItHappened() {
            public boolean itHappened() {
                return Files.contents( logFile ).contains( marker );
            }
        }, 1000 );

        //Did the exception stack trace appear in the log?
        assert !Files.contents( logFile ).contains( "Exception" );
        return true;
    }
```

This is a good example to show the importance of writing the test before we fix the bug. To be sure that the test is correct, we have to prove that the final assertion

```
assert !Files.contents( logFile ).contains( "Exception" );
```

fails before we fix the bug and passes after the bug has been fixed. If we had fixed the bug first, we could never be sure our test was detecting the error, and hence never be sure our bug fix had done what we wanted.

Remember to Roll!

When we are testing the contents of log files, it is essential that we clean up our log files between tests. For example, suppose that we have a test that a certain error message is produced in some situation. We write a test for it as follows:

```
public boolean errorIsLoggedTest() {
    init();
    //Check that the error message is not in the log file.
    ...
    //Cause the error.
    ...
    //Wait for and check the error message.
    ...
    cleanup();
    return true;
}
```

Now suppose that the same error message is to be produced in another test. If we don't clean up the log file between the new test and the old one, the old test will fail.

In logging systems such as that built into the JDK, a convenient way of cleaning up the logs is to re-initialize the logger so that it 'rolls' the log files. See the `FileLogger` class for details.

Testing What is Printed to the Console

There are times when our software prints messages to the command line that are important enough to need testing. An example might be messages printed during startup in some kind of diagnostic mode. Another example is any output from a command-line program such as GrandTestAuto (see Chapter 19).

In this section, we'll look at two different strategies for testing what has been printed to the console. The first is to set the value of `System.out` to be a stream that we can intercept and test. The second is to run our program in a separate JVM from which we read the standard output and error streams.

Logging

Switching Streams

Let's look at how the stream-switching approach can be used to test a simple `Assert` class.

Prior to JDK 1.4, an `Assert` class with its own `AssertException` was pretty much obligatory in any test system. With the arrival of the `assert` statement in Java 1.4, however, the requirements for a custom-built assertion system are much reduced. Our tests use an `Assert` class for testing object equality. The added value from this class, over just using `assert obj1.equals(obj2)`, is that the objects are printed out if they are not equal. This greatly simplifies debugging. If the objects are not equal, but their string versions are, the classes of the objects are printed out too. This avoids the confusion that occasionally arises when two objects that have the same `toString()`, such as a `StringBuffer` and a `String`, are compared.

To test our `Assert` class, we need to check not only that the comparison of different objects results in an exception, but also that the appropriate message is printed out. To do this, we set `System.out` to be a `PrintStream` wrapping a `ByteArrayOutputStream`. We can look at the data in the `ByteArrayOutputStream` to check what was printed to `System.out`. Here is the test:

```java
public boolean equalTest() throws IOException {
    //Check some objects that are equal.
    Assert.equal( "", "" );
    Assert.equal( "string", "string" );
    Assert.equal( Boolean.TRUE, Boolean.TRUE );
    Assert.equal( 123, 123 );

    //Now check some that are different.
    //Check that an error is thrown, and also what
    //is printed to System.out.
    ByteArrayOutputStream byteOut = new ByteArrayOutputStream();
    PrintStream newOut = new PrintStream( byteOut );
    PrintStream oldOut = System.out;
    System.setOut( newOut );

    boolean result = true;
    try {
        Assert.equal( "", null );
        result = false;//Should have got an exception
    } catch (Exception e) {
        //Ignore.
    }
    newOut.flush();
    String message = byteOut.toString();
    String expected = "1st Object: ''" + StringUtils.NL +
            "2nd Object: 'null'" + StringUtils.NL;
    result &= expected.equals( message );
```

```
        byteOut.reset();
        try {
            Assert.equal( "", "not blank" );
            result = false;//Should have got an exception
        } catch (Exception e) {
            //Ignore.
        }
        newOut.flush();
        message = byteOut.toString();
        expected = "1st Object: ''" + StringUtils.NL +
                   "2nd Object: 'not blank'" + StringUtils.NL;
        result &= expected.equals( message );
        byteOut.reset();
        try {
            Assert.equal( true, false );
            result = false;//Should have got an exception
        } catch (Exception e) {
            //Ignore.
        }
        newOut.flush();
        message = byteOut.toString();
        expected = "1st Object: 'true'" + StringUtils.NL +
                   "2nd Object: 'false'" + StringUtils.NL;
        result &= expected.equals( message );
        byteOut.reset();
        try {
            Assert.equal( "", new StringBuffer() );
            result = false;//Should have got an exception
        } catch (Exception e) {
            //Ignore.
        }
        newOut.flush();
        message = byteOut.toString();
        expected = "1st Object: ''" + StringUtils.NL +
                   "2nd Object: ''" + StringUtils.NL +
                   "Class of o1: class java.lang.String" + StringUtils.NL +
                   "Class of o2: class java.lang.StringBuffer"+
                                                    StringUtils.NL;
        result &= expected.equals( message );
        //Reset System.out.
        System.setOut( oldOut );
        return result;
    }
```

Note that this test does not throw any assertions or exceptions after `System.out` has been switched. This is so that the second last line, in which `System.out` is restored to its old value, is executed even if the test fails.

Reading the Output From a Second JVM

We'll now look at how we can test what is printed to the console by running our program in a separate JVM, from which we read System.err and System.out. The examples here are from the tests for GrandTestAuto.

One of the requirements of GrandTestAuto is that we can have the test results printed to the console. Obviously we need to test this requirement, the question is how?

Our solution is as follows. The test creates a new JVM in which GrandTestAuto is run on a configured set of test classes in the zip file Grandtestauto.test36_zip. The output of the new JVM is read, and the appropriate checks are made:

```java
public class LoggingToConsole extends FTBase {
    public boolean runTest() {
        //Run GTA for a configured package
        //in a separate JVM and read the console output.
        File classesZip = new File( Grandtestauto.test36_zip );
        String settingsName = Helpers.expandZipAndWriteSettingsFile(
                classesZip, true, true, true, null, true );
        String sout = Helpers.runGTAInSeparateJVM(
                                    settingsName )[1];
        //Check that the output contains:
        //...the settings summary
        assert sout.contains( new Settings().summary() ) : sout;
        //...the unit test results
        assert sout.contains( Messages.message(
                Messages.TPK_UNIT_TEST_PACK_RESULTS,
                "a36", Messages.passOrFail( true ) ) );
        //...the function test results
        assert sout.contains( Messages.message(
                Messages.TPK_FUNCTION_TEST_PACK_RESULTS,
                "a36.functiontest", Messages.passOrFail( true ) ) );
        //...the load test results
        assert sout.contains( Messages.message(
                Messages.TPK_LOAD_TEST_PACK_RESULTS,
                "a36.loadtest", Messages.passOrFail( true ) ) );
        //...and the overall result.
        assert sout.contains( Messages.message(
                Messages.OPK_OVERALL_GTA_RESULT,
                Messages.passOrFail( true ) ) );
        return true;
    }
}
```

Obviously, the `Helpers` class is doing most of the work here, with the method:

```java
public static String[] runGTAInSeparateJVM( String... args ) {
    StringBuilder cmd = new StringBuilder();
    cmd.append( "java.exe " );
    cmd.append( "-Duser.dir=\"" );
    cmd.append( System.getProperty( "user.dir" ) );
    cmd.append( "\"" );
    cmd.append( " -cp \"" );
    cmd.append( System.getProperty( "java.class.path" ) );
    cmd.append( "\" org.grandtestauto.GrandTestAuto" );
    for (String arg : args) {
        cmd.append( " " );
        cmd.append( arg );
    }
    final StringBuilder soutBuilder = new StringBuilder();
    final StringBuilder serrBuilder = new StringBuilder();
    try {
        Process p = Runtime.getRuntime().exec( cmd.toString() );
        BufferedReader sout = new BufferedReader(
                new InputStreamReader( p.getInputStream() ) );
        ProcessOutputReader soutReader =
                new ProcessOutputReader( sout, soutBuilder );
        soutReader.start();
        BufferedReader serr = new BufferedReader(
                new InputStreamReader( p.getErrorStream() ) );
        ProcessOutputReader serrReader =
                new ProcessOutputReader( serr, serrBuilder );
        serrReader.start();
        soutReader.join();
        serrReader.join();
    } catch (Exception e) {
        e.printStackTrace();
    }
    return new String[]{
            soutBuilder.toString(),
            serrBuilder.toString()};
}
```

`ProcessOutputReader` is a little `Thread` subclass for reading a stream into a `StringBuffer`:

```java
private static class ProcessOutputReader extends Thread {
    private final BufferedReader toRead;
    private final StringBuilder sb;
```

Logging

```java
        public ProcessOutputReader( BufferedReader sout,
                                    StringBuilder sb ) {
            this.toRead = sout;
            this.sb = sb;
        }
        public void run() {
            try {
                String line = toRead.readLine();
                while (line != null) {
                    sb.append( line );
                    sb.append( NL );
                    line = toRead.readLine();
                }
            } catch (Exception e) {
                try {
                    toRead.close();
                } catch (Exception e1) {
                    //Ignore.
                }
            }
        }
    }
```

Once we've built tools like this to run our software in a separate JVM, we can write all manner of useful tests. For example, there was a bug in GrandTestAuto by which the logger was not being flushed with each log event. This was a problem because when a test failed catastrophically and brought down the whole JVM, we were not sure which one it was, since the log file had not been updated. The fix for this bug was simply to flush the log with each event. Before we put the fix in, though, we wrote a test in which GrandTestAuto runs a set of configured tests that includes a call to `System.exit()` before the last test class is run:

```java
public boolean runTest() {
    testsRun.clear();
    //Run package 81 where the test class
    //a80.functiontest.D
    //calls System.exit( ).
    Helpers.runGTAInSeparateJVMAndReadSystemErr(
            new File( Grandtestauto.test81_zip ),
            true, true, true, null );
    String loggedToFile = Files.contents(
            new File( Settings.DEFAULT_LOG_FILE_NAME ) );
    //Check that test classes have been run even after the
    //test that called System.exit()
```

```
        assert loggedToFile.contains( "public a81.E()");
        assert loggedToFile.contains( "public a81.C()");
        //...rest of test follows
        return true;
}
```

This test, which requires a JVM exit, clearly is one that needs the program under test to be run in a separate JVM.

Summary

It is pretty straightforward to check the contents of log files and this is something we must do if our software has requirements about what is logged. We have presented two schemes for testing the output of console-based programs: switching `System.out` with a stream we can intercept and running our software in a separate JVM from which we can read `System.out` and `System.err`.

14
Communication with External Systems

Over the years we've had to develop a number of interfaces between LabWizard and external applications. Mostly, these have been data channels between our software and laboratory information systems or analytical instruments, and these are too specific for discussion here. However, we have also used email as a communication device, spreadsheets for data mining, and PDF as a report format. As a result we have come up with some techniques for testing these aspects of our software.

In this chapter, we will discuss two different techniques for testing that our use of email is bug-free. We will then turn to Excel spreadsheets and PDF documents. After this, we look at some of the problems that can occur when we serialize our objects, and how we can test for them. Finally, we demonstrate some tools that make interactions with the file system easier to test.

Email

LabWizard relies on email for a number of different tasks. In early versions of the software, email was used only to send log files to our support facility. Then the LabWizard server was enhanced to send emails to report critical errors. These emails go to a paging service where they are converted to phone text messages directed to support engineers. The messages contain details of the site and the error, and often an engineer is able to fix the problem before the customer is inconvenienced. Our most recent application of email is to allow rules-based messages from a LabWizard Knowledge Base. For example, a pathologist might write a rule that sends a glucose result directly to a patient and their doctor.

These kinds of applications require tests that first provide the trigger for our system to send email messages, and then check that the required emails were actually sent. We will present here two approaches to checking email functionality.

The code we discuss here is in the packages `jet.testtools` and `jet.testtools.test`.

Using an External Email Account

The first approach is to have an email address with our email service provider to which emails are sent during tests. When starting a test, all messages are removed from the mailbox of the remote account. The test code that triggers the email is run, and then we run the code that waits for the expected message to be received.

The class `MailBox`, shown next, simplifies these kinds of tests. The important methods are `removeAllMessages()` and `findMessagesMatching(String subject)`:

```java
/**
 * Simple wrapper for a POP3 inbox.
 */
public class MailBox {
    private String host;
    private String user;
    private String password;
    public MailBox( String host, String user, String password ) {
        this.host = host;
        this.user = user;
        this.password = password;
    }
    public void removeAllMessages() throws Exception {
        iterateMessages( new WorkForMessage() {
            public void doIt( Message message ) {
                try {
                    message.setFlag( Flags.Flag.DELETED, true );
                } catch (MessagingException e) {
                    e.printStackTrace();
                }
            }
        } );
    }
    public Set<Message> getMessagesMatching(
                        final String subject ) throws Exception {
        final Set<Message> result = new HashSet<Message>();
```

```
        iterateMessages( new WorkForMessage() {
            public void doIt( Message message ) {
                try {
                    if (subject.equals( message.getSubject() )) {
                        result.add( message );
                    }
                } catch (MessagingException e) {
                    e.printStackTrace();
                }
            }
        } );
        return result;
    }
    private interface WorkForMessage {
        void doIt( Message message );
    }
    private void iterateMessages(
                WorkForMessage wfm ) throws Exception {
        Properties properties = new Properties();
        Session session = Session.getInstance( properties, null );
        Store store = session.getStore( "pop3" );
        store.connect( host, user, password );
        Folder inbox = store.getFolder( "INBOX" );
        inbox.open( Folder.READ_WRITE );
        Message[] allMessages = inbox.getMessages();
        for (Message message : allMessages) {
            wfm.doIt( message );
        }
        inbox.close( true );
        store.close();
    }
}
```

For example, here is a unit test of our example Server using a remote email account:

```
/**
 * Test of the mail functionality using a
 * remote mail server.
 */
public boolean remoteEmailTest() throws Exception {
    init( false );//Use remote mail account and create mailBox.
    mailBox.removeAllMessages();
    //Send a support email.
    final String subject = "Support message from site x";
```

```
    String body = "Test message only";
    server.sendSupportEmail( subject, body );
    //Wait for up to a minute for the expected message to arrive.
    boolean gotIt = Waiting.waitFor( new Waiting.ItHappened() {
        public boolean itHappened() {
            try {
                return mailBox.getMessagesMatching(
                        subject ).size() > 0;
            } catch (Exception e) {
                e.printStackTrace();
                return false;
            }
        }
    }, 60000, 5000 );
    assert gotIt : "Did not get expected message!";
    //Now check the message details.
    Set<Message> messages = mailBox.getMessagesMatching( subject );
    assert messages.size() == 1;
    //etc
    cleanup();
    return true;
}
```

As this example shows, checking email functionality with a remote account is a pretty easy thing to do. However, there are a few pitfalls.

The most significant issue is that the mailbox that is being tested is a resource shared by all the instances of our testing system. So, if we are running builds and tests on multiple machines, we can get test errors caused by one test machine deleting the emails sent by others. Of course, if only one machine is running tests, this is not an issue. If multiple machines are running tests, it can be worked around by providing a mailbox per machine.

A second problem is that tests with a real email account are slower than those with a local account. How significant this problem is depends on how many email-related tests we are running.

Another issue is that remote mailboxes are not completely reliable, so these tests need to be engineered to try several times if there are connection problems

Finally, if our test machine has no internet connection, we cannot use a remote email account in our tests. It's unlikely that a dedicated test machine would be offline. What often happens for the authors, though, is that they are writing and testing code while on the train or otherwise disconnected from the internet.

Using a Local Email Server

The difficulties and performance issues with using a remote email account in our tests mean that sometimes it is more convenient to use an email server running inside our test JVM. The class under test will connect to this email server, and we can make programmatic calls on the email server itself to check for the expected messages.

There are several freely available email servers written in Java that we can either use directly or modify for our requirements. We have adapted **Eric Daugherty's JES** software to build a class, `SMTPProcessor`, which acts as a simple email server.

To use `SMTPProcessor` in our `ServerTest`, we need to set up our `Server` instance to use an email server available on the localhost. This is done in `init()`, where we also create and start the `SMTPProcessor`:

```
private void init() throws Exception {
    ...
    //The server uses a local email server.
    String emailHost = InetAddress.getLocalHost().getHostAddress();
    server = new Server( inputDir, outputDir, archiveDir,
                emailHost, "harry@hogwarts.edu" );
    //Create and start the mail server.
    mailServer = new SMTPProcessor( false );//No debug.
    mailServer.start();
    ...
}
```

Although the act of sending the message itself is by a call in our test thread, the mail server has its own internal thread listening for and processing connections, so in our test we still need to wait for the message to be received:

```
/**
 * Test of the mail functionality using an email server
 * running inside the test JVM.
 */
public boolean sendSupportEmailTest() throws Exception {
    init( true );

    //send a support email.
    String subject = "Support message from site x";
    String body = "Test message only";
    server.sendSupportEmail( subject, body );

    //Wait for the expected message to arrive.
    Waiting.waitFor( new Waiting.ItHappened() {
        public boolean itHappened() {
            return mailServer.messagesReceived().size() == 1;
```

```
            }
    }, 1000 );
    //Now check the message details.
    Collection<SMTPMessage> messages =
                            mailServer.messagesReceived();
    assert messages.size() == 1;
    SMTPMessage message = messages.iterator().next();
    Assert.equal( message.getFromAddress().toString(),
            "harry@hogwarts.edu" );
    //etc.
    cleanup();
    return true;
}
```

This test runs more quickly than the version using the remote server.

Which Method is Best?

Testing email functionality using an email server running in the same JVM as the test program is certainly more convenient than relying on a remote email server, and the tests run more quickly. On the other hand, in our deployed software we need the messages to be relayed by different systems without, for example, being rejected as spam. In this sense, the approach using a remote email server is more valid. A compromise that we adopted to get the best from both approaches is to use a local email server for unit tests and a remote email server for function tests where test execution speed is less important.

Testing Spreadsheets

LabWizard produces a lot of different data extractions as Excel spreadsheets since this is a format that our customers are very familiar with and is one to which they can apply their own analysis tools. We use a package called JExcel (see http://jexcelapi.sourceforge.net) for reading and writing Excel files. There are quite a few Java packages for working with Excel and there is also a Java API for Sun's Star Office suite. Whatever package we use, the principles for testing import and export functionality will be the same.

In our tests we use a complementary pair of methods found in the TestHelper class:

```
String[][] spreadsheetToArray(File spreadSheetFile, int sheetIndex);
```

and

```
void arrayToSpreadsheet(File spreadSheetFile, int sheetIndex,
                                    String[][] dataforSheet);
```

Suppose that we want to test a method, `writeToSpreadsheet()`, which creates a spreadsheet file and inserts some string values into cells in the first sheet. Our test will read the resulting spreadsheet and compare the retrieved values with the expected values.

```java
public boolean writeToSpreadSheetTest() throws Exception {
    //write the spreadsheet using the production method.
    File spreadSheetFile = writeToSpreadsheet();
    //check whether the retrieved values of the
    //spreadsheet are as expected.
    int sheetIndex = 0;
    String[][] retrieved = TestHelper.spreadsheetToArray(
                    spreadSheetFile, sheetIndex );
    String[][] expected = new String[][]{
                    {"00", "01"},
                    {"10", "11"}};
    //Check the retrieved data.
    checkData( expected, retrieved );
    return true;
}
```

The `checkData()` method is:

```java
private static void checkData( String[][] expected,
                    String[][] retrieved ) {
    //Check the number of rows.
    assert retrieved.length == expected.length;
    for (int row = 0; row < expected.length; row++) {
        //Check the number of columns.
        assert retrieved[row].length == expected[row].length;
        for (int col = 0; col < expected[row].length; col++) {
            //Check the data in each cell.
            assert retrieved[row][col].equals( expected[row][col] );
        }
    }
}
```

A test for a method that reads a spreadsheet will be the converse:

```java
public boolean readFromSpreadSheetTest() throws Exception {
    //Write test data into a spreadsheet.
    String[][] expected = new String[][]{
                        {"00", "01"},
                        {"10", "11"}};
    File spreadSheetFile = new File( "Spreadsheet.xls" );
    int sheetIndex = 0;
    TestHelper.arrayToSpreadsheet(
            spreadSheetFile, sheetIndex, expected );
    //Read the spreadsheet using the production method.
```

```
            String[][] retrieved = readFromSpreadSheet( spreadSheetFile );
            //Check that the retrieved values of the spreadsheet
            //are as expected.
            checkData( expected, retrieved );
            return true;
        }
```

The pair of helper methods, `spreadsheetToArray()` and `arrayToSpreadsheet()`, can be readily generalized to handle data types other than strings.

PDF

Adobe Portable Document Format (PDF) is the standard for producing documents that are rendered on the screen and on a printer in pretty much the same way. With LabWizard, it is possible to generate reports as PDF documents, and we use a package called iText (see http://www.lowagie.com/iText/) for this.

In our tests we want to check that the documents we generate contain the text we expect them to. We may also want to check the formatting, but the content is most critical. Extracting the text from a PDF document using iText is surprisingly difficult, for reasons explained in the iText documentation. A good tool for the job is the pdfbox package (see http://www.pdfbox.org). Using these libraries, it is possible to create complementary reader and writer methods for PDF, in much the same way as with spreadsheets.

To read the contents of a PDF we use this method from `jet.testtools.TestHelper`:

```
    public static String pdfFileContents( File pdf ) throws Exception {
        //Use the PDFBox tool to write the text to a file.
        long now = System.currentTimeMillis();
        String tempFileName = "PDFTemp" + now + ".txt";
        File temp = new File( Files.tempDir(), tempFileName );
        org.pdfbox.ExtractText.main( new String[]{
                            pdf.getAbsolutePath(),
                        temp.getAbsolutePath()} );
        //Use the Apache commons-io tool to read the
        //text file, then deleted it.
        String result = FileUtils.readFileToString( temp );
        temp.delete();
        return result;
    }
```

Once we have this test helper method, it is easy to check whether we have generated the correct content of a PDF file. In the following example from a LabWizard function test, we start an Administrator client, create some user accounts, generate a PDF report of these accounts, and check with a configured PDF report:

```
public boolean pdfReportTest() {
    //Start administrator client
    AdminTestProxy admin = TestBase.startAdminClient();
    //Create a set of user accounts
    //......
    //Generate a user account report using the Administrator menu
    File file = new File( TestSetup.tempDirectory(), "report.pdf" );
    admin.userAndGroupReport( file );
    //Read the contests of the report
    String found = TestHelper.pdfFileContents( file );
    //Read the contents of a configured report
    File expectedPdfFile = new File( TestData.ExpectedReport_pdf );
    String expectedPdf =
                TestSetup.pdfFileContents( expectedPdfFile );
    //Ignore formatting differences by stripping whitespace
    //before the comparison
    Assert.aequals(StringUtils.stripWhitespace( found ),
            StringUtils.stripWhitespace( expectedPdf ) );
    return true;
}
```

We have found one annoying thing about the PDFBox package, which is an undocumented reliance on another package called FontBox, see www.fontbox.org.

Serialization

A class is made serializable simply by declaring it to implement java.io.Serializable, which is a marker interface. (That is, it defines no methods.) Because we get serializability almost for free, it is easy to forget to test it, and there are several ways in which it can fail.

Our most recent LabWizard serialization failure was as follows. We had a class encapsulating the results of a fairly serious server-side calculation. The result objects were serialized and sent to the client programs, from where they could be saved as spreadsheets. We had a function test for this process, but the test crashed. The underlying problem was that although our results class implemented Serializable, it had an inner class that did not.

To test that our classes really are Serializable, we can use SerializationTester:

```
/**
 * For testing whether instances of
 * <code>java.io.Serializable</code> really are
 * serializable. This code is taken from a JavaWorld
 * article about deep copying:
 * "Java Tip 76: An alternative to the deep copy technique"
```

```
 * By Dave Miller, JavaWorld.com, 08/06/99.
 * See http://www.javaworld.com/javaworld/javatips/jw-javatip76.html
 */
public class SerializationTester {
    public static boolean check( Serializable object ) {
        boolean result = false;
        ObjectOutputStream oos = null;
        ObjectInputStream ois = null;
        try {
            ByteArrayOutputStream bos = new ByteArrayOutputStream();
            oos = new ObjectOutputStream( bos );
            oos.writeObject( object );
            oos.flush();
            ByteArrayInputStream bin =
                    new ByteArrayInputStream( bos.toByteArray() );
            ois = new ObjectInputStream( bin );
            Object deserialized = ois.readObject();
            result = deserialized != null;
        } catch (Exception e) {
            e.printStackTrace();
            result = false;
        } finally {
            try {
                oos.close();
            } catch (Exception e) {
                //Don't worry.
            }
            try {
                ois.close();
            } catch (Exception e) {
                //Don't worry.
            }
        }
        return result;
    }
    //Prevent instantiation.
    private SerializationTester() {
    }
}
```

The unit test for this class is quite instructive in that it demonstrates the serialization failures that occur:

```
public class SerializationTesterTest {
    public boolean checkTest() {
        //Check that something that really
        //is Serializable passes.
        assert SerializationTester.check(
                new IsSerializable() );
        //Check that something that is not
```

```
            //Serializable fails. This will print an exception.
            System.out.println(
                    "****** EXPECTED STACK TRACE BEGINS ******" );
            assert !SerializationTester.check(
                    new NotActuallySerializable() );
            System.out.println(
                    "****** EXPECTED STACK TRACE ENDS ******" );
            return true;
        }
    }
    class IsSerializable implements Serializable {
        private Thing thing = new Thing();
        public int value() {
            return thing.value;
        }
        private class Thing implements Serializable {
            int value;
        }
    }
    class NotActuallySerializable implements Serializable {
        private Thing thing = new Thing();
        public int value() {
            return thing.value;
        }
        private class Thing {
            int value;
        }
    }
```

The difference between the classes `IsSerializable` and `NotActuallySerializable` is their inner class. In `IsSerializable`, the inner class is serializable, in `NotActuallySerializable` it is not.

Files

If our software reads or writes files then our test code will probably involve a fair amount of file management, if only to copy configured data to locations from which it can be used in tests. To reduce the drudgery of writing these tests, it may well be worth having a utility class with static methods for the most commonly used operations. For example, the tests of the code in this book use a class, `Files`, for this purpose. We first saw this class in Chapter 3, where it was introduced in the context of test infrastructure, namely where to put temporary files created in tests. As we saw in Chapter 3, this class has methods for temporary file management that are slightly different from those provided by `java.io.File`. The relevant methods are:

```
    public static File tempDir();
```

which provides a standard directory for tests to write files in, and

```
public static File cleanedTempDir();
```

which provides the same directory but first removes all files and sub-directories.

Another set of file convenience methods is for the handling of zip archives. We have

```
public static void zip( File[] filesToZip, File zippedFile );
```

and its inverse:

```
public static void unzip( String zipName, File destination );
```

plus a method for comparing archives:

```
public static boolean equalZips( File zip1,
                                 File zip2 ) throws Exception;
```

For further details, look at the source code for `jet.testtools.Files` and also the unit tests.

Other common operations in testing are:

- Copying files to a directory.
- Removing all of the files from a directory.
- Reading the contents of a file as a single `String`.
- Comparing the contents of two files.
- Checking the time at which a file was last changed.

For these, and lots of other tasks, we suggest using the `FileUtils` class in the Apache Commons IO package, `org.apache.commons.io`.

Summary

In this chapter we developed two approaches to testing our email functionality. We've also seen how to test the reading and writing of Excel spreadsheets and PDF documents. We also looked at a method for checking that the classes we declare to implement `java.io.Serializable` really are serializable. Finally, we discussed tools for file management in tests.

15
Embedding User Interface Components in Server-side Classes

In our development of LabWizard, we've found a couple of design patterns to be really invaluable in terms of making the software simpler and easier to test. One of these is the use of handler interfaces, which we looked at in Chapter 6. In the present chapter, we'll look at the advantages of another well-known design pattern, which is making what are commonly thought of as server-side classes responsible for their own display.

Suppose that we are writing software in which there is the concept of a User. We will need to consider:

- How User objects are persisted.
- The runtime behavior of User objects.
- The presentation of User objects on screen.

It is a fairly common practice for these three aspects of a business concept such as User to be distributed amongst two or three classes. The persistence and business logic aspects are thought of as belonging in 'server-side' classes, and the user interface is written as a 'client-side' class. There can be good reasons for making this separation of concerns (we discuss these later), but if adopted indiscriminately it can lead to unnecessary coupling between classes, a proliferation of 'setter' methods, and unnecessary testing. For an argument against slavish adoption of the Model-View-Controller pattern where it is not warranted, see Allen Holub's article *Build user interfaces for object-oriented systems, Part 2: The visual-proxy architecture*, www.javaworld.com/javaworld/jw-09-1999/jw-09-toolbox.html.

A Typical MVC System

Consider a class `User` comprising of a name and a password. A simple implementation might be:

```
public class User {
    private String name;
    private String password;
    public void setName( String name ) {
        this.name = name;
    }
    public String name() {
        return name;
    }
    public void setPassword( String password ) {
        this.password = password;
    }
    public String password() {
        return password;
    }
}
```

On the server-side we want to store the collection of all users in a class called `UserManager`, for example:

```
public class UserManager {
    private Set<User> users = new HashSet<User>();
    public void addUser( User user ) {
        users.add( user );
    }
    //etc...
}
```

On the client-side we want to be able to create new `User` instances using a GUI. The client application has a main class `Client` with a **Manage users** menu and an **Add new user** menu item, as shown below:

For the sake of simplicity, we ignore other facilities such as showing the list of existing users, modifying the properties of a selected user, removing a user, and so on.

When the menu item is activated, Client creates an instance of an AddUser class that provides the GUI components for an **Add new user** dialog. It then creates a User instance once the name and password fields have been completed and the **Add user** button has been pressed. The following figure shows the user interface for creating a new user:

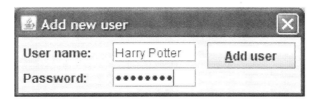

A minimal implementation of Client might be:

```
public class Client {
    private JFrame frame = new JFrame();
    private Client() {
        final AddUser.Handler handler = new AddUser.Handler() {
            public JFrame frame() {
                return frame;
            }
            public void addUser( User user ) {
                //lookup the remote server e.g. via RMI
                UserManager userManager = new UserManager();
                userManager.addUser( user );
            }
        };
        Action addUserAction = new AbstractAction( "Add new user..." )
        {
            public void actionPerformed( ActionEvent e ) {
                new AddUser( handler ).show();
            }
        };
        JMenu userMenu = new JMenu( "Manage users" );
        userMenu.add( new JMenuItem( addUserAction ) );
        JMenuBar menuBar = new JMenuBar();
        menuBar.add( userMenu );
        frame.setTitle( "Client application" );
        frame.setJMenuBar( menuBar );
```

```
            frame.pack();
            frame.setVisible( true );
        }
        public static void main( String[] args ) {
            UI.runInEventThread( new Runnable() {
                public void run() {
                    new Client();
                }
            } );
        }
    }
```

The AddUser class might be implemented as:

```
    public class AddUser {
        private Handler handler;
        public interface Handler {
            JFrame frame();
            void addUser( User user );
        }
        public AddUser( Handler handler ) {
            this.handler = handler;
        }
        public void show() {
            final JDialog dialog = new JDialog( handler.frame(),
                    "Add new user", false );
            final JTextField nameField = new JTextField( 7 );
            final JTextField passwordField = new JPasswordField( 7 );
            JButton button = new JButton( "Add user" );
            button.setMnemonic( 'a' );
            button.addActionListener( new ActionListener() {
                public void actionPerformed( ActionEvent e ) {
                    User user = new User();
                    user.setName( nameField.getText() );
                    user.setPassword( passwordField.getText() );
                    handler.addUser( user );
                    dialog.dispose();
                }
            } );
            JPanel panel = new JPanel( new GridLayout( 2, 2, 5, 5 ) );
            panel.setBorder( BorderFactory.createEmptyBorder
                                            ( 5, 5, 5, 5 ) );
            JLabel nameLabel = new JLabel( "User name:" );
```

```
            JLabel passwordLabel = new JLabel( "Password:" );
            panel.add( nameLabel );
            panel.add( nameField );
            panel.add( passwordLabel );
            panel.add( passwordField );
            Box box = Box.createVerticalBox();
            box.setBorder( BorderFactory.createEmptyBorder
                                              ( 5, 5, 5, 5 ) );
            box.add( button );
            box.add( Box.createVerticalGlue() );
            dialog.getContentPane().add( panel, BorderLayout.CENTER );
            dialog.getContentPane().add( box, BorderLayout.EAST );
            dialog.pack();
            dialog.setVisible( true );
        }
    }
```

Note the `Handler` interface required in the `AddUser` constructor. An implementation of this is created by `Client` when constructing an `AddUser` instance. Apart from providing a `JFrame` so that the dialog has an owner, the handler provides the `void addUser( User user )` method. This is, in effect, a callback method that provides notification when the **Add user** button has been pressed. The `Client` (or at least its inner class implementing `AddUser.Handler`) can then make a call on the server's `UserManager.addUser( User user )` method to add the new user to the collection.

The Problem

So what is wrong with this design? Lots!

The first problem is that in order for the `AddUser` class to set the values of a `User` object when calling the `Handler.addUser(User user)` method, the `User` class has to effectively expose its fields by providing the public `setName()` and `setPassword()` methods. The same access would be needed for the GUI components that would enable us to edit the properties of a user. And of course once we have exposed these methods, they have to be explicitly tested. Furthermore, any new property that we would want to associate with a user would need to be implemented by both `User` and `AddUser`.

Similarly, in order to construct the menu items on the **Manage users** menu, `Client` needs to know the user management facilities that are available. This is unnecessary complexity for `Client`, resulting in more testing. And again, any new user management facility that is required will need to be implemented in both `UserManager` to provide the new server-side functionality and `Client` to provide the corresponding new GUI components, an unnecessary maintenance overhead.

In other words, the design is flawed because of the tight coupling between `UserManager` and `Client` and between `User` and `AddUser`.

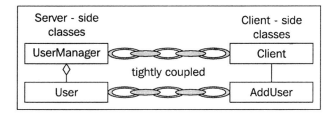

The Solution

To avoid this tight coupling and the overly-exposed fields that it implies, let us refactor `User` and `UserManager` to make them responsible for their own GUI components, rather than saddling the `Client` class with this responsibility.

```java
public class UserV2 {
    private String name;
    private String password;
    public String name() {
        return name;
    }
    public String password() {
        return password;
    }
    public static UserV2 newUser( JFrame frame ) {
        final JDialog dialog = new JDialog
                          ( frame, "Add new user",  false );
        final JTextField nameField = new JTextField( 7 );
        final JTextField passwordField = new JPasswordField( 7 );
        final UserV2 user = new UserV2();
        JButton button = new JButton( "Add user" );
        button.setMnemonic( 'a' );
        button.addActionListener( new ActionListener() {
            public void actionPerformed( ActionEvent e ) {
                user.name = nameField.getText();
                user.password = passwordField.getText();
                dialog.dispose();
            }
        } );
```

```java
                JPanel panel = new JPanel( new GridLayout( 2, 2, 5, 5 ) );
                panel.setBorder( BorderFactory.createEmptyBorder
                                                    ( 5, 5, 5, 5 ) );
                JLabel nameLabel = new JLabel( "User name:" );
                JLabel passwordLabel = new JLabel( "Password:" );
                panel.add( nameLabel );
                panel.add( nameField );
                panel.add( passwordLabel );
                panel.add( passwordField );
                Box box = Box.createVerticalBox();
                box.setBorder( BorderFactory.createEmptyBorder
                                                    ( 5, 5, 5, 5 ) );
                box.add( button );
                box.add( Box.createVerticalGlue() );
                dialog.getContentPane().add( panel, BorderLayout.CENTER );
                dialog.getContentPane().add( box, BorderLayout.EAST );
                dialog.pack();
                dialog.setVisible( true );
                return user;
            }
        }
```

UserV2 can create an instance of itself using a GUI with the newUser() method. The former User methods setName() and setPassword() methods are gone, as well as the AddUser class itself. The client application no longer has to know anything about the internals of a user.

But we are not finished yet. We continue this refactoring approach with UserManagerV2.

```java
        public class UserManagerV2 {
            private Set<UserV2> users = new HashSet<UserV2>();
            public JMenu menu( final JFrame frame ) {
                Action addUserAction = new AbstractAction( "Add new user..." )
{
                    public void actionPerformed( ActionEvent e ) {
                        users.add( UserV2.newUser( frame ) );
                    }
                };
                JMenu userMenu = new JMenu( "Manage users" );
                userMenu.add( new JMenuItem( addUserAction ) );
                return userMenu;
            }
            //methods for persisting the collection of users, etc...
        }
```

Given a `JFrame`, `UserManagerV2` provides a menu that our new version of the client component, `ClientV2`, can put into its main window. Of course, this would not work if the `userManager` was a remote object, but some variation of it would. `ClientV2` does not have to be aware of what is on the menu, or even what needs to be done if a menu item is activated—all this functionality has been encapsulated within `UserManagerV2`.

Compared with the former `Client`, the implementation of `ClientV2` is now almost trivial:

```
public class ClientV2 {
    private JFrame frame = new JFrame();
    public UserManagerV2 userManager;

    public ClientV2() {
        //Lookup the remote server's UserManager e.g. via RMI
        userManager = new UserManagerV2();

        JMenuBar menuBar = new JMenuBar();
        menuBar.add( userManager.menu( frame ) );

        frame.setTitle( "Client application" );
        frame.setJMenuBar( menuBar );
        frame.pack();
        frame.setVisible( true );
    }
    public static void main( String[] args ) {
        UI.runInEventThread( new Runnable() {
            public void run() {
                new ClientV2();
            }
        } );
    }
}
```

Our new design has removed the tight coupling between the server-side classes and the client-side classes:

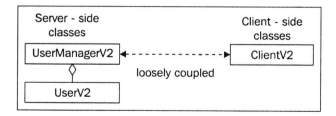

Not only are the refactored classes more maintainable, but the testing task is much easier—there are fewer public methods to test and fewer classes to test. Moreover, testing of some operations, such as adding a user, is localized to the particular class providing that functionality, rather than being split into server-side tests and client-side tests for the operation.

Which Approach Should We Use?

We've seen that there are several advantages to a design in which our server-side objects can produce their own user interfaces:

- The classes do not have to expose fields for the user interface class to update.
- The code is shorter and simpler.
- There are fewer methods to test.

On the other hand, there are scenarios where separating user interface classes from our server-side objects is a more appropriate design. Suppose, for example, we are working on a distributed system in which the client software is connected via a low bandwidth connection to the server. In this situation, we may want to send a bare minimum of data to the client, and this might mean using cut-down versions of our business objects on the client side.

As another example, consider again a distributed system in which we have proprietary algorithms in our server-side classes. We might be reluctant for the client programs to contain the code for these algorithms, as it is very hard to protect Java bytecode, even with obfuscation tools. Therefore, our client software would use some kind of cut-down versions of the server-side classes.

Summary

We should consider keeping user interface components integrated into the class containing the corresponding data members and business logic:

Extreme Testing Guideline: We should not necessarily separate what should be a single class into a client-side class containing user interface components and a server-side class containing the data members.

If we can adhere to this guideline, our maintenance and testing jobs will be much easier.

16
Tests Involving Databases

One of the difficulties with unit testing complex systems is dealing with databases. If our software interacts with one or more databases — either to make queries on an external system or as a persistence mechanism — we will need to test our software using databases in some known state.

One way of handling this problem is to use a tool such as DbUnit (see www.dbunit.org) to populate databases with pre-determined data sets. Such databases can then be used in unit tests. Apart from being restricted to relational databases, a problem with this approach is that being "database-centric", it does not easily mesh with the "class-centric" approach for unit testing. For example, if the database schema changes then not only do our production classes and test classes have to change, but also the DbUnit scripts providing the data sets.

An approach which avoids these issues is to build an abstraction layer in our software that hides the details of persistence from our higher level application classes, and at the same time facilitates testing. This is the approach we have taken in our development of LabWizard, and we will describe it in this chapter. Using this approach, it will be easy for us to test the following:

- Is our data in fact being persisted?
- Does our software behave as expected when using specific configurations of stored data?
- Is the new version of our software compatible with existing databases deployed at our customer sites?
- If we have more than one database implementation, can we migrate from one to another?
- Can our software detect and recover from database corruptions?

Moreover, our approach makes it very easy to change database vendors or even to change database technologies. For example, from a relational database to a B-Tree database.

A Uniform Approach to Accessing the Database

Whether we use a relational database, an object-oriented database, or some special-purpose database like a B-Tree database, it is important to have a consistent, high-level, and as far as possible, transparent way of persisting our objects. What we have to avoid is the maintenance and testing nightmare of each class using low-level database calls to persist its own data.

The area of persistence design is a complex one, and there are many packages that hide the underlying database mechanisms to a greater or lesser extent. We outline here an approach we have found suitable for LabWizard, where the primary function of the database is simply to provide a persistent object store for that application. This approach may not be suitable where there are further requirements for the database to be used for ad-hoc queries or reports, or directly interfaced to other applications.

We first define a façade called `Database` which wraps the low-level database transactions appropriate for the particular database implementation being used. In LabWizard we have both a relational implementation of `Database` and a B-Tree implementation. Our current relational version uses Microsoft Jet databases, though in the past we have used a number of pure Java databases, at least at the evaluation stage. We persist with Jet as it provides a convenient tool, Microsoft Access, for us to directly examine or edit our databases on those very rare occasions when there are inconsistencies in the stored data. For our deployed software, we always use a high-performance B-Tree implementation which is based on the Infinity Database Engine (see `http://www.infinitydb.com`).

The details of these implementations do not concern us here — suffice to say that a `Database` can store and retrieve mementos that represent the state of our persisted objects. An instance of `Database` is associated with a single file, and the database's name is the root of the corresponding file name.

With this approach, databases can be conveniently managed by a `DatabaseManager` defined as follows:

```
public class DatabaseManager {
    /** Obtain the unique instance of this class. */
    public static DatabaseManager instance() {
        ...
    }
    /** Opens or creates a database with the specified name. */
    public Database open( String name ) {
        ...
    }
```

```java
    /** Closes the database with the specified name. */
    public void close( String name ) {
        ...
    }
    /** Deletes the database with the specified name. */
    public void delete( String name ) {
        ...
    }
    /** Deletes all databases */
    public void deleteAll() {
        ...
    }
    /** Names of all databases, open or closed. */
    public Set<String> names() {
        ...
    }
    /** True if the specified database is open. */
    public boolean isOpen( String name ) {
        ...
    }
    /**
     * The directory from where all databases can be opened.
     */
    public File databaseDirectory() {
        ...
    }
    /**
     * The file corresponding to the specified database name.
     */
    public File fileForDatabase( String name ) {
        ...
    }
    /**
     * True if the corresponding database is not corrupt.
     */
    public boolean checkDB( String databaseName ) {
        ...
    }
}
```

A class whose instances need to be persisted will extend `Storable`:

```
public abstract class Storable {
    protected abstract String stringSerialize();
    protected abstract void restore( String str );
    ...
}
```

A container managing instances of these classes will extend or delegate to `Manager`:

```
public class Manager<T extends Storable> {
    public Manager( Database database ) {
        ...
    }
    public void put( T storable ) {
        ...
    }
    public void remove( T storable ) {
        ...
    }
    public Set<T> getAll() {
        ...
    }
    //Plus many other convenience methods...
}
```

In summary, the objects we store extend `Storable`, and these are stored in a `Database` by a `Manager`:

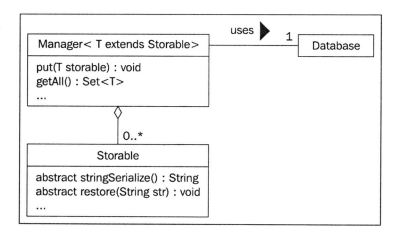

Each `Manager` uses its reference to a `Database` to store and retrieve `Storable` instances. Once we have implementations of `Database`, `DatabaseManager`, `Storable`, and `Manager`, the rest is easy.

Suppose we wish to persist instances of `User` (see Chapter 15). We do this by making `User` extend `Storable`, and `UserManager` contain a `Manager`:

```java
public class UserManager {
    private Manager<User> users;
    public UserManager(Database database) {
        users = new Manager<User>(database);
    }
    public void addUser(User user) {
        users.put(user);
    }
    public Set<User> allUsers() {
        return users.getAll();
    }
    public int size() {
        return users.size();
    }
    //plus many other methods specific to managing Users...
}
```

`UserManager` delegates to `Manager`, rather than extends it, as there are only these few `Manager` methods needed.

`User` will be responsible for serializing itself to a `String` by implementing the abstract `stringSerialize()` method of `Storable`. Similarly, it will be responsible for restoring itself from a `String` by implementing the abstract `restore(String)` method. We use an XML representation of the object's data, as this format is simple to implement in Java using DOM, but any convenient serialization format can be used.

A major advantage of making `User` extend `Storable` and being responsible for its own serialization is that its database "schema" is private. No other class needs to know which fields in `User` need to be persisted, or how these fields are serialized. The implementation of `User` is independent of the database technology used. In fact, `User` can be thoroughly unit tested without having to instantiate a database at all. The exception of course is regression tests, where we need to check that instances of `User` can be restored by itself from configured databases.

Persistence Testing

With this design, basic persistence testing consists of two steps. Firstly, we check the serialization and de-serialization methods of the persisted class. For example, in the unit test for User we have:

```
public boolean stringSerializeTest(){
    //Create a user with a known name and password.
    //Use a test extension of User that simply allows us
    //to set these fields without a GUI
    User user = new UserExt("My user name", "My password");
    String str = user.stringSerialize();
    User restoredUser = new User();
    restoredUser.restore(str);
    Assert.equal(user, restoredUser);
    return true;
}
```

Secondly, we check by reconstructing our manager from the database that it is storing these objects persistently. For example, in the unit test for UserManager we have:

```
public boolean addUserTest() {
    //Delete any databases from previous tests
    init();
    User user = new UserExt("My user name", "My password");
    Set<User> expectedUsers = new HashSet<User>();
    DatabaseManager dbManager = DatabaseManager.instance();
    Database db = dbManager.open("MyDatabaseName");
    UserManager userManager = new UserManager(db);
    //check whether the UserManager is initially empty
    Assert.equal(userManager.allUsers(), expectedUsers);

    //check whether the UserManager contains an added user
    userManager.addUser(user);
    expectedUsers.add(user);
    Assert.equal(userManager.allUsers(), expectedUsers);

    //check whether UserManager is persisting its users by
    //verifying whether a reconstructed UserManager still
    //contains that user
    userManager = new UserManager(db);
    Assert.equal(userManager.allUsers(), expectedUsers);

    //cleanup
    dbManager.deleteAll();
    return true;
}
```

Both in the init() and in the final cleanup we use a DatabaseManager method to simply delete all databases known to the DatabaseManager.

Database Management

In addition to the simple persistence tests, there are many other aspects of database management that need to be tested. For example, as part of a regression test, we want to check that existing databases are still compatible with the latest version of our code. To do this, we first copy a configured database file from our test data location (see Chapter 3) and then attempt to access the corresponding database:

```
public boolean databaseRegressionTest() {
    final String databaseName = "Configured";
    Files.copyConfiguredDB(databaseName);
    DatabaseManager dbManager = DatabaseManager.instance();
    Database db = dbManager.open(databaseName);
    //Can we open the configured database?
    UserManager userManager = new UserManager(db);
    //Can we retrieve the all the users?
    int expectedNumberOfUsers = 5000;
    Assert.equal(userManager.size(), expectedNumberOfUsers);
    //further tests...
    //cleanup
    dbManager.deleteAll();
    return true;
}
```

If we have several implementations of Database, we need to check that we can convert between these implementations. In the following example this is done by the class DatabaseConverter which can convert our Infinity B-Tree database into a relational database and vice versa.

```
public boolean databaseConversionTest() {
    DatabaseManager dbManager = DatabaseManager.instance();
    String databaseName = "Configured";
    Files.copyConfiguredDB(databaseName);
    Database dbInf = dbManager.open(databaseName);
    //Convert the Infinity database to an equivalent
    //Relational database
    Database dbRel = DatabaseConverter.toRelational(dbInf);
    //Convert back to Infinity
    Database doubleConverted = DatabaseConverter.toInfinity(dbRel);
    //Check whether the double conversion has produced a database
    //which is identical to the original
    assert doubleConverted.identicals(dbInf);
    //cleanup
    dbManager.deleteAll();
    return true;
}
```

Tests Involving Databases

Hardware being as it is, it is a fact of life that the content of our databases may become corrupt. We need to check that our software can detect and recover from database corruptions. The following example shows a simple test to check that when a corrupt database is opened, an exception is thrown and a corresponding entry is recorded in the log. (In fact a corrupt LabWizard database results in an email that is converted to a text message directed to the support team.)

```java
public boolean openCorruptDatabaseTest() {
    DatabaseManager dbManager = DatabaseManager.instance();
    String databaseName = "CorruptDBName";
    Files.copyConfiguredDB(databaseName);
    //Check that the database file is indeed corrupt.
    assert !dbManager.checkDB(databaseName);
    //Roll the log to clear it.
    FileLogger.rollLogOver();
    //Attempt to open the database. This is expected
    //to fail and generate a log entry.
    boolean ok = false;
    try {
        dbManager.open(databaseName);
    } catch (Exception e) {
        //expected exception
        ok = true;
    }
    //Is there an entry in the log?
    final String expectedLogEntry = "Corrupt database: "
                                        + databaseName;
    final File logFile = new File(
                FileLogger.logDirName, "log.txt.0");
    final Waiting.ItHappened itHappened = new Waiting.ItHappened() {
    public boolean itHappened() {
            return Files.contents(logFile
                                    ).contains(expectedLogEntry);
        }
    };
    assert Waiting.waitFor(itHappened, 1000);
    //Cleanup.
    dbManager.deleteAll();
    return ok;
}
```

This is of course not a comprehensive list of database tests, but it does give a pretty good idea of the kinds of database testing that is easily done using our database management design.

Summary

By wrapping all low-level database transactions in a `Database` class and by providing a `DatabaseManager` class, we have insulated our application classes from any specific database code. This allows us to easily swap database vendors, even changing from relational to B-Tree databases. There are further advantages in testing as we can add, delete, and copy configured database files with ease.

17
Function Tests

As we discussed in Chapter 1, LabWizard is tested in three phases:

- Unit tests of individual classes.
- Function tests of complete applications.
- Load tests.

In Chapters 4 to 11, we developed techniques for unit testing user interfaces. In Chapters 12 to 16, we concentrated on the unit testing of server-side classes. To write function tests, we need to make use of the tools and techniques from these two strands. In this chapter, we will show how this can be done, using some LabWizard function tests as examples. The examples we have chosen run LabWizard clients and the LabWizard server in separate JVMs and simulate a laboratory information system, so they are relatively complex tests.

Before delving into the examples, we will look at how function tests are specified and documented.

Specification of the Tests

Function tests come from user requirements, typically documented in a Software Requirements Specification or by XP "use cases".

For each requirement we design one or more test cases in our test specification. For each test case, we write a Java class that actually runs the test.

This might all seem very abstract and academic, so let's see how it actually applies to the development of LabWizard.

LabWizard can be deployed as what we call the Clinical Reporting Application, which consists of four major components:

- A **Knowledge Builder**, which allows the expert user to build and maintain Knowledge Bases.
- A **Server**, which interfaces to an online information system presenting cases to the Knowledge Bases to be interpreted.
- A **Validator**, which allows users to review interpretations before they are sent back to the laboratory information system.
- An **Administrator**, which allows an IT user to manage the LabWizard user accounts, Knowledge Bases, and data interfaces.

The LabWizard Clinical Reporting Application is defined by the documentation and code hierarchy shown below:

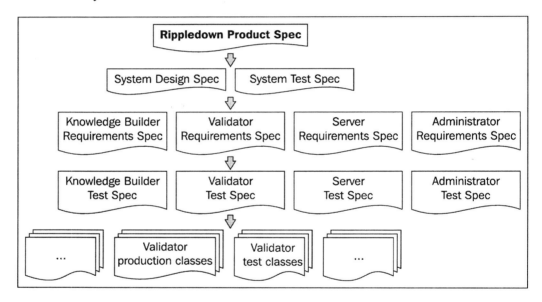

The **Rippledown Product Spec** at the top describes in very general terms the structure of the product as a server connected to the specific client programs. Requirements that apply to all client programs, such as the need for users to log in, are stated here. For requirements specific to the individual clients, the reader is referred to the corresponding client program requirements spec. By the way, the term **Rippledown** refers to the rule building algorithm on which LabWizard is based.

The **System Design Spec** gives a rough architectural overview of the software. This specifies, for example, that client/server communication is via RMI. For more detailed design information, the System Design Spec refers the reader to design documents for each of the client programs (not shown here). The **System Test Spec** describes the levels of testing we do, namely unit, function, and load testing, and as expected, refers the reader to the individual product test specifications.

The next level of documents are the requirements specs for the client programs. It is at this level that the bulk of the software requirements are documented. Here is an example of one of these requirements:

> **2.1.11   delete**
> The Validator shall provide the facility for the User to delete the currently showing case.
>
> Refines   Rippledown Product Spec.Validate Interpretations
> ImplementedBy   rippledown.val
> TestedBy   Validator Test Spec.DeleteCase

The structure of this requirement is as follows. After the statement of requirement, the **Refines** heading indicates a reference to the originating requirement in the Rippledown Product Specification. The **ImplementedBy** heading indicates a reference to the Java package `rippledown.val` implementing this requirement. Finally, the **TestedBy** heading indicates a hyperlinked reference to a test called **DeleteCase** in the Validator Test Specification. This test is shown below:

> **4.2.31   DeleteCase**
> *Check whether the user is able to delete the currently showing case*
> 1. Start Rippledown server, and add a project in pre-validate mode.
> 2. Send 2 cases from the online information system simulator.
> 3. Start Validator client and start reviewing the project queue.
> 4. Use the menu item to delete the current case.
> 5. Check that the number of remaining cases is now 1.
> 6. Delete the current case again.
> 7. Check that the number of remaining cases is now 0.
> 8. Check that after 5 seconds, no cases have been sent back to the online information system simulator.
> 9. Exit the validator client.
> 10. Restart the Rippledoen server.
> 11. Restart the Validator client.
> 12. Check that there are still no cases on the project queue.

Function Tests

This test starts the LabWizard Server and sends two cases to the Knowledge Base using a simulator. These cases are queued to the Validator so that their interpretations can be reviewed by an expert. Once the Validator is started in Step 3 in the previous figure, its main window shows the number of queued cases. The following figure shows the **LabWizard Validator**, showing that two cases are waiting to be reviewed for the Knowledge Base **AProject**:

The user selects the **AProject** queue and clicks the **Review** button. The **Validator Case Viewer** dialog is then displayed, enabling the user to review the cases on the queue, one after another.

In this test however, the user deletes the cases. The test then checks the following:

- When the user deletes a case, does the system indicate that the number of remaining cases is decremented? (Steps 5 and 7.)
- When the user deletes a case, is there an unwanted side-effect whereby the case is actually sent back to the online information system? (Step 8.)
- When the user deletes a case, is the case deleted from the database, as evidenced by checking the state of the system after a server restart? (Step 12.)

Implementation of the 'DeleteCase' Test

To implement this test, we create an empty class `DeleteCase`. As discussed in Chapter 3, this class shall be located in the package `rippledown.val.functiontest`. Next, we use each step of the test specification as a code comment. Finally, we implement each step with one or more calls to our high-level test helper classes. In this example, the test helper classes are:

- `TestBase`: For starting the server and any of the clients.
- `ServerRunner`: Allows access to the Server.
- `KBTestProxy`: A UI Wrapper for the Knowledge Builder.
- `AdminTestProxy`: A UI Wrapper for the Administrator.
- `ValTestProxy`: A UI Wrapper for the Validator main screen.
- `UIValidatorCase`: A UI Wrapper for the Case Viewer.
- `LISTestProxy`: Simulates an online information system.
- `Assert`: An assertion class that pre-dates the one in Java itself, and still has some useful methods, including `aequals()`, which shows the differences between the compared objects if they are not identical.

Recall that the UI Wrapper classes implement mouse and keyboard actions using `Cyborg`. They also check what is displayed by finding user interface components within the frames of the test JVM, and reading their state in a thread-safe fashion. It's important to note that no 'backdoor' access to the clients is required. With these helper classes, the implementation of the test class follows very simply from its specification:

```
public class DeleteCase implements AutoLoadFunctionTest {
    public boolean runTest() {
        //1. Start LabWizard Server and add
        //a project in pre-validate mode.
        ServerRunner serverRunner = TestBase.startServer();
```

Function Tests

```
KBTestProxy kb =
    TestBase.startKBClientAndCreateOnlineDefaultProject();
kb.exit();
AdminTestProxy admin =
        TestBase.startAdminClientAndDefaultPreProject();
admin.exit();

//2. Send 2 cases from
//the online information system simulator.
LISTestProxy simulator =
                InterfaceFactory.createLISTestProxy();
simulator.provideMultiRFI( 2 );

//3. Start Validator client and start reviewing the project.
ValTestProxy val = TestBase.startValidatorClient();
val.waitForCases( 2 );
val.selectAndStartQueueBySequence( TestBase.PROJECT_A );
UIValidatorCase uiCase = val.getUICase();

//4. Delete the current case.
uiCase.waitForApprovedButtonClickable();
uiCase.deleteCase();

//5. Check that the number of remaining cases is now 1.
val.waitForRemainingCasesToBeExactly( 1 );

//6. Delete the current case again.
uiCase.waitForApprovedButtonClickable();
uiCase.deleteCase();

//7. Check that the number of remaining cases is now 0.
val.waitForRemainingCasesToBeExactly( 0 );

//8. Check that after 5 seconds, no cases have been sent
//to the online information system simulator.
TestSetup.pause( 5 );
Assert.aequals( simulator.getNumberOfInterpretations(), 0 );

//9. Exit the Validator client
uiCase.back();
val.exit(); //Note: this does not kill the Validator's JVM

//10. Restart the LabWizard server.
try {
    serverRunner.stop();
} catch (RemoteException e) {
    //Guaranteed to happen as the JVM exits. Ignore.
}
TestBase.restartServer();
```

```
            //11. Restart the Validator client.
            val = TestBase.startValidatorClient();
            //12. Check that there are no cases on the project queue.
            val.waitForRemainingCasesToBeExactly( 0 );
            TestBase.cleanupUseSparingly();
            return true;
        }
    }
```

`DeleteCase` implements `AutoLoadFunctionTest` so that it can be executed automatically using GrandTestAuto (more on this in Chapter 19).

Even using high-level helper classes, function test classes tend to be fairly lengthy, as a single step in the test specification may sometimes require several lines of code to implement. It's far easier to implement several smaller tests than a single large test that attempts to test several requirements. To minimize test complexity and execution time, we recommend the following:

Extreme Testing Guideline: Design each test case around a single requirement, or for a complex requirement, just a single aspect of that requirement.

Function test execution times can also be lengthy in comparison with unit testing. For example, the `DeleteCase` test takes just under a minute. To minimize test execution time:

Extreme Testing Guideline: Don't write function tests for what can adequately be tested at the unit level.

In the Validator, for example, there is a requirement for the delete menu to be disabled if there is no case currently displayed. Rather than design a function test for this, we design a unit test of the user interface class which manages the menus and the display of the case. There is no need to involve the server or other LabWizard clients in a lengthier function test.

When designing a test specification up front, we can be completely focused on what needs to be tested and what doesn't. However, when we shift gears and implement the tests, we have to be focused on the details of getting them to execute, which can be quite messy. At that time, we really need to follow the plan laid out by the code comments, which are taken from our test specification.

Extreme Testing Guideline: Always write the test specification before implementing the test class, and use the specification as code comments.

Tests Involving Multiple JVMs

The simplest function tests will have GUI components of the application executing in the JVM of the test program. In the `DeleteCase` test above, the Validator, Knowledge Builder, and Administrator clients all execute in the test JVM as this simplifies the use of the `Cyborg` for key presses, mouse movements, and the reading of the state of the GUI controls. One of the limitations of this approach, however, is that the clients can never be allowed to truly exit or else the test JVM would also exit. In the example above, the call `uiClient.exit()` in step 9 disposes the client's frame, stops any timers, and so on, but does not call `System.exit()`.

In contrast, we require the LabWizard Server to execute in its own JVM during tests as we invariably need to check whether the state of the system is persisted in the Server's database and is restored after a Server restart or crash. Also, as our client-server architecture uses RMI, it is important that our function tests also use this mechanism. Occasionally, this uncovers a serialization bug or some other flaw that was not found in the unit tests.

So how does the test start the Server in its own JVM, and once started, how does the test get the Server to stop and restart as in step 10 above?

The approach we use is for the test to start a new JVM using `Runtime.getRuntime().exec()`, just as we started new JVMs in the GrandTestAuto function test in Chapter 13. The main class that the JVM starts is `ServerRunnerImpl`. This class implements a `Remote` interface named `ServerRunner` that has the methods we need for controlling the LabWizard Server:

```java
public interface ServerRunner extends Remote {
  /**
   * Request that the Server stop itself.
   */
  void stop() throws RemoteException;
  /**
   * Shut down the server JVM without calling stop on the server.
   */
  void shutdownJVM() throws RemoteException;
  //Other server commands
  //...
}
```

`ServerRunnerImpl` has its own LabWizard Server on which it executes the commands needed by test classes, such as stopping the server in step 10 of `DeleteCase`. Note that this method also causes the Server's JVM to exit:

```java
public class ServerRunnerImpl extends UnicastRemoteObject
                              implements ServerRunner {
    private RippledownServer rippledownServer;
    public static void main( String[] args ) {
        //Details omitted.
    }
    private ServerRunnerImpl() {
        //Create RippledownServer.
        //Details omitted.
    }
    public void stop() throws RemoteException {
        rippledownserver.shutdown( "" );
        System.exit( 0 );
    }
    //Other server commands
    //...
}
```

The `main()` method of `ServerRunnerImpl` creates and exports a `ServerRunnerImpl` via RMI:

```java
public class ServerRunnerImpl extends UnicastRemoteObject
                              implements ServerRunner {
    ...
    public static void main( String[] args ) {
        ServerRunnerImpl runner = new ServerRunnerImpl();
        try {
            Naming.rebind( REGISTERED_NAME, runner );
            //Print to std out that we've registered with RMI.
            //The listening process will know
            //then that it can connect.
            System.out.println( REGISTERED_ANNOUNCEMENT );
        } catch (Exception e) {
            System.err.println( "Runner not registered: " +
                                            REGISTERED_NAME );
            e.printStackTrace();
        }
    }
    ...
}
```

Function Tests

Note the message printed to System.out indicating a successful creation and export. The class that started the JVM captures and parses this stream, and when it finds this message it attempts to connect to the ServerRunner using RMI. This creation of a ServerRunnerImpl in a separate JVM is in a convenience method in one of our test infrastructure classes, TestBase:

```
ServerRunner runner TestBase.startServer();
```

In summary, function tests such as DeleteCase have a remote reference to an object that has a direct reference to a server as shown next:

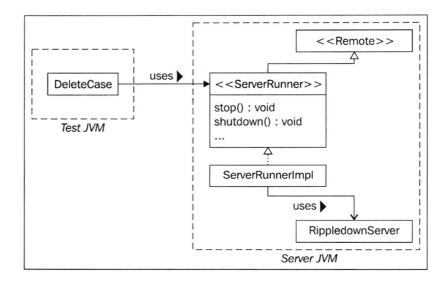

Multiple JVMs with GUI Components

We can extend the methodology for having the server in its own JVM to implement function tests which specify that one of the client programs, for example the Validator, should also execute in a separate JVM. See steps 6 and 8 next:

> **4.2.14 RecoveryAfterUnexpectedTermination**
> *Check that no cases are held up by a client crash. STR 2219.*
> 1. Start the Rippledown Server.
> 2. Start a Knowledge Builder client and create project A.
> 3. Start an Administrator client.
> 4. Add the project as a Validator project.
> 5. Send 20 cases for the project.
> 6. Start a Validator client for project A in a separate JVM.
> 7. Commence reviewing the project, but don't approve any cases.
> 8. Terminate the separate Validator JVM.
> 9. Start a new Validator client.
> 10. Select project A and commence reviewing the project.
> 11. Check that there are 20 cases on the queue by approving them all and checking that they are sent to the Online Information System.

This test was designed to reproduce a bug documented in a **Software Trouble Report** (STR). The reported bug was that if a Validator client crashed, for example due to a power failure, once the user restarted the Validator they could no longer review all the required cases.

Once we start the Validator client in its own JVM, that is, separate to both the test JVM and the server JVM, how do we read its state and control it with key presses and mouse clicks as required by step 7 shown previously?

The answer is to create a remote implementation of our Validator UI Wrapper class, and access this remote instance from our test JVM using RMI. The interface `RemoteValTestProxy` specifies a remote version of the Validator UI Wrapper class `ValTestProxy`:

```
public interface RemoteValTestProxy extends Remote {
    /**
     * Select the named queue.
     */
    void selectQueue( String queueName ) throws RemoteException;
    /**
     * Start the currently selected queue.
     *
     * @return A case viewer proxy for reviewing the cases.
     */
    RemoteUIValidatorCase startQueue() throws RemoteException;
    /**
     * Exit the JVM without shutting down the Validator
```

Function Tests

```
    */
    void shutdownJVM() throws RemoteException;
    //Other methods.
    //...
}
```

`RemoteValTestProxyImpl` is an implementation of `RemoteValTestProxy` which simply delegates these methods to our non-remote helper class `ValTextProxy`:

```
public class RemoteValTestProxyImpl extends
                UnicastRemoteObject implements RemoteValTestProxy {
    private ValTestProxy valTestProxy;
    public RemoteValTestProxyImpl( ValTestProxy valTestProxy )
                                        throws RemoteException {
        this.valTestProxy = valTestProxy;
    }
    public void selectQueue( String queueName )
                                        throws RemoteException {
        valTestProxy.selectQueue( queueName, "" );
    }
    public RemoteUIValidatorCase startQueue()
                                        throws RemoteException {
        valTestProxy.startQueueBySequence();
        return new RemoteUIValidatorCaseImpl(
                            valTestProxy.getUICase() );
    }
    public void shutdownJVM() throws RemoteException {
        System.exit( -1 );
    }
    //Other Validator commands
    //...
}
```

The `startQueue()` method returns a `RemoteUIValidatorCase`, a remote version of `UIValidatorCase`, which is a UI Wrapper for the Validator Case Viewer.

To implement our function test, we define a test class `RecoveryAfterUnexpectedTermination` that has a remote reference to `RemoteValTestProxyImpl` and uses this to implement steps 6, 7, and 8 shown previously.

This diagram shows the relation between the test JVM, the server JVM, and the Validator JVM:

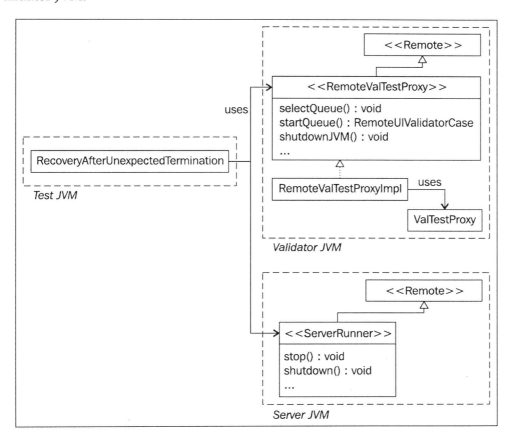

With these remote helper classes, the implementation of the test class again follows very simply from its specification:

```
public class RecoveryAfterUnexpectedTermination
                    implements AutoLoadFunctionTest {

    public boolean runTest() throws Exception {
        //1. Start the Rippledown Server.
        TestBase.startServer();

        //2. Start Knowledge Builder client and create project A.
        KBTestProxy kbProxy = TestBase.
                startKBClientAndCreateOnlineDefaultProject();
        kbProxy.exit();

        //3. Start an Administrator client.
        //4. Add the project as a Validator project.
```

Function Tests

```
            AdminTestProxy adminProxy =
                    TestBase.startAdminClientAndDefaultPreProject();
            adminProxy.exit();
            //5. Send 20 cases for the project
            LISTestProxy lis = InterfaceFactory.createLISTestProxy();
            lis.provideMultiRFI( 20 );
            //6. Start a Validator client for
            //project A in a separate JVM
            RemoteValTestProxy remoteValTestProxy =
                            TestBase.startRemoteValidatorClient();
            //7. Commence reviewing the project,
            //but don't approve any cases.
            remoteValTestProxy.selectQueue(
                                    TestBase.TEST_PROJECT_NAME );
            RemoteUIValidatorCase remoteValCase =
                                    remoteValTestProxy.startQueue();
            remoteValCase.waitForApprovedButtonClickable();
            //8. Terminate the separate Validator JVM.
            try {
                remoteValTestProxy.shutdownJVM();
            } catch (Exception e) {
                //Guaranteed as the remote JVM is shutting down. Ignore.
            }
            //9. Start a new Validator client.
            ValTestProxy valQueue = TestBase.startValidatorClient(
                                    "Porky", "Frankenstein" );
            //10. Select project A and commence reviewing the project.
            valQueue.selectAndStartQueueBySequence(
                                    TestBase.TEST_PROJECT_NAME );
            UIValidatorCase vCase = valQueue.getUICase();
            //11. Check that there are 20 cases on the queue
            //by approving them all and checking that they are
            //sent to the Online Information System.
            vCase.approveMultiCasesWithoutChange( 20 );
            vCase.back();
            valQueue.exit();
            lis.waitForExactlyTheSpecifiedNumberOfInterpretations( 20 );
            lis.cleanup();

            //cleanup remote objects
            try {
                UnicastRemoteObject.unexportObject(remoteValCase, true );
            } catch (NoSuchObjectException e) {
                //Ignore.
            }
            TestBase.cleanupUseSparingly();
            return true;
        }
    }
```

Use of a Function Test as a Tutorial

The testing infrastructure developed to provide full automation of our function tests can also be used to implement tutorials. Both a function test and a tutorial specify a sequence of user interactions. In the former, we check whether the system behaves as expected. In the latter, we simply want to prompt the user to follow the specified actions as an exercise. So how does our testing infrastructure help here?

Firstly, the application's user interface may change from release to release. Even a simple change such as moving a menu item from one menu to another can introduce an error into the specification of our tutorial. How do we find these errors? If we use our test helper classes to automatically run the tutorial as though it were a function test, we can check that each user interaction has the expected result. If the tutorial has specified a user interaction that is no longer valid, this will show up as a test failure. Hence running the tutorial in this way is providing a regression test.

Secondly, whilst some tutorials will start with a completely empty system, most will start with a populated database that provides a more meaningful context for the sequence of user operations to be performed. How do we set up the system to be in a configured state before the tutorial starts? Again, before we prompt the user to start the tutorial, we can use our test helper classes to set up the system with a configured database, or alternatively, to automatically run through a preliminary sequence of user actions.

The following example is the specification of a LabWizard Knowledge Builder tutorial that gives the user practice in creating a rule using the syntax `at least`, which is a key phrase in the LabWizard rule condition language:

Using 'at least' in a condition

Before you start this tutorial, please wait for the project 'AtLeast' to be opened in the Knowledge Builder. There will be 14 cases on the rejected case list.

Steps:

1. Find and select the case 450008.
2. Note that the interpretation of this case is "Raised fasting glucose".
3. Note also that there is a previous episode for this patient where the GTT 2hr test value is 11.3 which is above the diabetic limit of 11.1.
4. Build a rule to add the conclusion " Consistent with diabetes mellitus" to this interpretation under the report section heading "GTT Conclusion".
5. Use the condition "At least 1 GTT 2hr >= 11.1"
6. Note that the interpretation is now "Raised fasting glucose. Consistent with diabetes mellitus."

Function Tests

The specification is maintained in the file KBAtLeast.html so that we can easily associate it with its implementation class. In addition to the TestBase and KBTestProxy helper classes that we've already met, we use a class called HelpShower for showing the tutorial steps to the user.

```java
public class KBAtLeast extends Tutorial {
    public boolean runTutorial() throws Exception {
        //Before you start this tutorial,
        //please wait for the project
        //'AtLeast' to be opened in the Knowledge Builder.
        //There will be 14 cases on the rejected case list.
        String dbName = "AtLeast";
        TestBase.startServerForTutorial();
        TestBase.copyTrainingProjectToServerDirectory( dbName );
        KBTestProxy kb = TestBase.startKBClient();
        kb.project().open( dbName );
        kb.wCL().waitForCaseListSize( 14 );

        if (runningAsATutorial) {
            ProcessControl pc =
                    HelpShower.showTutorialSteps( getClass() );
            waitForEndOfUserSteps();
            pc.destroySubProcess();
            return true;
        }
        //Steps:
        //1. Find and select the case 450008.
        String caseName = "450008";
        kb.wCL().find( caseName );
        kb.cView().waitForCaseInCaseView( caseName );

        //2. Note that the interpretation of this case is
        //   "Raised fasting glucose".
        String oldComment = "Raised Fasting Glucose.";
        kb.focusToReportView();
        Assert.aequals( kb.reportView().report(), oldComment );

        //3. Note also that there is a previous episode
        //for this patient  where the GTT 2hr test value is 11.3,
        //which is above the diabetic limit of 11.1.
        Assert.aequals( kb.cView().numberOfEpisodes(), 2 );
        String found = kb.cView().valueForAttribute( "GTT 2hr", 0 );
        Assert.aequals( found, "11.3" );

        //4. Build a rule to add the conclusion
        //   "Consistent with diabetes mellitus."
```

```
            //to this interpretation under the report
            //section heading "GTT Conclusion".
            String newComment = "Consistent with diabetes mellitus.";
            //5. Use the condition "At least 1 GTT 2hr >= 11.1"
            String condition = "At least 1 GTT 2hr >= 11.1";
            kb.rule().addConclusionWithExistingName(
                    "GTT Conclusion", newComment, condition );
            //6. Note that the interpretation is now
            // "Raised fasting glucose.
            //Consistent with diabetes mellitus."
            kb.focusToReportView();
            Assert.aequals( kb.reportView().report(),
                    oldComment + " " + newComment );
            TestBase.cleanupUseSparingly();
            return true;
        }
        /**
         * Execute as a tutorial rather than as a Function Test
         */
        public static void main( String[] args ) {
            runningAsATutorial = true;
            try {
                new KBAtLeast().runTutorial();
            } catch (Exception e) {
                e.printStackTrace();
            }
            TestBase.cleanupUseSparingly();
        }
    }
```

The method `KBAtLeast.main()` is called when the user picks this tutorial from the list of available tutorials. The `runTutorial()` method first sets up the tutorial using a configured database, starts the Server and Knowledge Builder client, and opens the project. As the flag `runningAsATutorial` has been set to true, the block of code:

```
    if (runningAsATutorial) {
        ProcessControl pc = HelpShower.showTutorialSteps(getClass());
        waitForEndOfUserSteps();
        pc.destroySubProcess();
        return true;
    }
```

Function Tests

shows a JavaHelp browser with a help page for the tutorial steps that the user is required to follow. The method `HelpShower.showTutorialSteps(getClass())` uses the fact that the HTML specification of the tutorial has the same root file name as the implementation class. Execution then pauses until the user has completed the steps manually. Here are the screens showing when the tutorial is running from our tutorial launcher application:

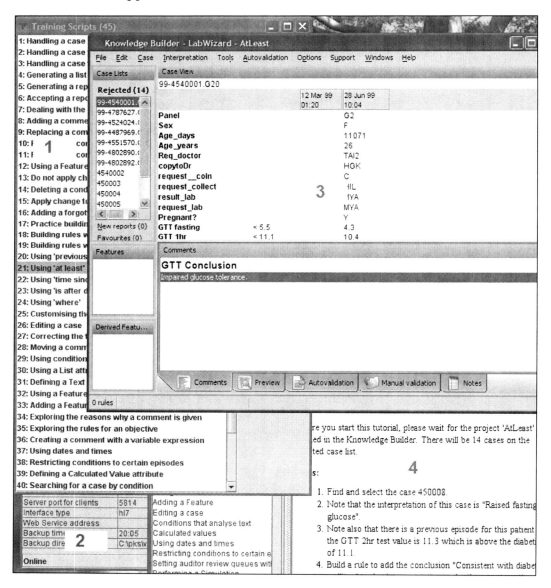

The screen **1** is our training scripts launcher. When item **21, Using 'at least'** is double-clicked, the Server, screen **2**, is launched in its own JVM. The server has been set up so that the configured Knowledge Base called **AtLeast** is available. Once the server is running, a **Knowledge Builder** client, screen **3**, is started and the **AtLeast** Knowledge Base is opened. Finally, a Help window showing the text of the tutorial, screen **4**, is launched. Incidentally, the Help system is shown in its own JVM to resolve some focusing issues and so that if the user closes the Help window, the launcher is not shut down.

When the method `KBAtLeast.runTutorial()` is run as a function test, the flag `runningAsTutorial` is set to false and the block of code that launches the help system and pauses is skipped. The remainder of the method simulates the user actions specified by the tutorial, checking the state of the system after each step to ensure that the steps are still valid.

The superclass `Tutorial` of `KBAtLeast` implements `AutoLoadFunctionTest`, so the test is run by GrandTestAuto automatically.

Testing a Web Service

LabWizard provides a web service that an external application can access according to our published WSDL (Web Service Description Language). In particular, an external application implemented in Java will typically generate its accessor classes using the Java utility `wsimport`, which is available in JDK 1.6. In keeping with our extreme testing methodology, our function tests should access the service in the same way. That is, our function tests should use accessor classes dynamically generated from the latest version of our WSDL. In this way, our tests are more likely to detect any inadvertent changes made to the WSDL.

However, this leads to the following "bootstrap" problem. We can't compile our function tests which include these accessor classes until the service has published its WSDL—but we can't publish our WSDL until our application is compiled and executing!

The solution is to build and publish our WSDL after compiling our production classes, but before compiling our test classes. So our build process is:

1. Compile the production classes.
2. Rebuild the WSDL using `wsgen`.
3. Execute the application and publish the WSDL.
4. Generate the test accessor Java source files from the WSDL using `wsimport`.
5. Stop the application.
6. Compile the test classes.

Function Tests

In step 1, we compile just our production classes. In step 2 we use the JDK utility `wsgen` to rebuild our WSDL from those production classes that provide the web service. In step 3, we execute this service and publish the new WSDL. Whilst the service is running, we can apply the Java utility `wsimport` to the WSDL in order to automatically generate the source code for our test accessor classes (step 4). We can then stop publishing the service (step 5) and, finally, compile all our test classes (step 6). This build sequence can be readily automated using Ant.

Let's demonstrate this build process using the web service provided by the `Request`, `Result`, and `Hello` classes located in the package `jet.webservice`:

Here is the `Request` class:

```
package jet.webservice;
public class Request {
    String name;
    public String getName() {
        return name;
    }
    public void setName( String name ) {
        this.name = name;
    }
}
```

and here is the `Response` class:

```
package jet.webservice;
public class Response {
    String message;
    public String getMessage() {
        return message;
    }
    public void setMessage( String message ) {
        this.message = message;
    }
}
```

A `Request` simply has a `name` field, and a `Response` a `message` field.

The web service is implemented by the `Hello` class. The single web method of this service, `query()`, creates a `Response` with a specified `message` field by simply prepending "Hello" to the `name` field retrieved from the `Request`. The `Hello` class also has a `main()` method which we shall look at later.

```java
package jet.webservice;
@WebService
public class Hello {
    public static final String WEB_SERVICE_ADDRESS =
            "http://localhost:8090/Hello?WSDL";
    private static String classDir = "../classes";
    private static String srcDir = "../src";
    private static String jawsTestPackage =
                                "jet.webservice.test.jaxws";
    private Endpoint endpoint;
    @WebMethod
    public Response query( Request request ) {
        Response result = new Response();
        result.setMessage( "Hello " + request.getName() );
        return result;
    }
    public Hello() {
        endpoint = Endpoint.publish( WEB_SERVICE_ADDRESS, this );
    }
    public void stop() {
        endpoint.stop();
    }
    ...
}
```

Steps 1 to 5 above are performed using the Ant task `build-hello`:

```xml
<target name="build-hello" depends="compile-nontest">
    <mkdir dir="${jaxws}"/>
    <exec executable="wsgen">
        <arg line="-cp ${classes} -verbose -r ${jaxws} -s ${src}
                        -d ${classes} -wsdl jet.webservice.Hello"/>
    </exec>

    <java classname="jet.webservice.Hello" fork="true" dir="${src}">
        <classpath refid="classpath" />
    </java>
</target>
```

Note that this task depends only on "nontest" classes. This allows all the test classes to be built subsequently in step 6.

After the subdirectory `jaxws` is created, the execution of the `wsgen` utility reads the `Hello` endpoint class and generates the WSDL and other required classes for web service deployment. These are located in the package `jet.webservice.jaxws`. The following figure shows the generated web service package:

The next step of the Ant task is to execute `main()` of the `Hello` class. This method publishes the endpoint (which is done in the `Hello` constructor), executes the `wsimport` utility to generate the Java classes corresponding to the published WSDL and then stops the endpoint. Alternatively, the steps done here in `main()` could be implemented as further Ant tasks.

```java
@WebService
public class Hello {
    //Rest of class listed above.
    public static void main( String[] args ) {
        Hello hello = new Hello();
        try {
            Process process = getRuntime().exec(
                    "wsimport -d " + classDir + " -s " + srcDir +
                    " -p " + jawsTestPackage +
                    " -verbose " + WEB_SERVICE_ADDRESS );
            showOutputFrom( process );
            process.waitFor();
        } catch (Exception e) {
            e.printStackTrace();
        }
        hello.stop();
    }
    private static void showOutputFrom( Process process )
            throws IOException {
        InputStream is = process.getInputStream();
        BufferedReader reader = new BufferedReader(
                new InputStreamReader( is ) );
        String line = reader.readLine();
        while (line != null) {
            System.out.println( line );
            line = reader.readLine();
        }
    }
}
```

Chapter 17

The classes generated by wsimport are created in a test package jet.webservice.test.jaxws.

These generated classes allow us to construct a "client view" of the web service, which is precisely what we need for our function tests (once we have completed our unit testing of the jet.webservice package of course!).

An example of a function test is specified by the SendRequest test case:

> **5.1.1 SendRequest**
> Check whether the query() method of Hello web service is able to be accessed
> 1. Start the Hello service.
> 2. Construct a Request generated from the WSDL.
> 3. Set the name in the Request to be "John Smith".
> 4. Make a query on the service with this Request.
> 5. Check that the message in the Response is "Hello John Smith"

Apart from publishing the Hello service itself in step 1, the test is implemented using the classes generated by the WSDL itself, as follows:

```
public class SendRequest implements AutoLoadTest {
    public boolean runTest() throws Exception {
        //1. Start the Hello service
        Hello hello = new Hello();

        //2. Construct a Request generated from the WSDL.
        jet.webservice.test.jaxws.Request request = new Request();

        //3. Set the name in the Request to be "John Smith"
        request.setName( "John Smith" );

        //4. Make a query on the service with this Request
        jet.webservice.test.jaxws.Hello proxyHello =
                    new HelloService().getHelloPort();
```

[255]

```
            jet.webservice.test.jaxws.Response response =
                        proxyHello.query( request );
        //5. Check that the message in the
        //Response is "Hello John Smith"
        Assert.equal( response.getMessage(), "Hello John Smith" );
        //Cleanup.
        hello.stop();
        return true;
    }
}
```

Summary

Function tests are those tests of an entire application that verify that our software implements an extreme programming 'use case', or some other specification of requirements.

Traditionally, these tests are thought of as being difficult to automate because of the large number of components involved and the high level at which the requirements are stated. However, by building infrastructure to run our server software in its own JVM, and by using the UI wrapper classes introduced in Chapter 7, the implementation of these tests is relatively straightforward.

For function tests of web services, we have shown how to build the test classes for a web service from its published WSDL. This ensures that the function tests are truly representative of the service we are testing, even as that service evolves.

18
Load Testing

Load testing is one area of software engineering where we believe that it can pay more to do less. We can put it no better than Donald E. Knuth and C. A. R. Hoare:

> "There is no doubt that the grail of efficiency leads to abuse. Programmers waste enormous amounts of time thinking about, or worrying about, the speed of noncritical parts of their programs, and these attempts at efficiency actually have a strong negative impact when debugging and maintenance are considered. We should forget about small efficiencies, say about 97% of the time: premature optimization is the root of all evil." (From "*Structured Programming with go to Statements*" by Donald E. Knuth, Stanford University, Stanford, California 94805, available from http://pplab.snu.ac.kr/courses/adv_pl05/papers/p261-knuth.pdf. Knuth is apparently quoting Hoare in the last sentence.)

The load tests we have developed for LabWizard are surprisingly few in number, and very simple in design. They either perform some critical operation a number of times and check that it does not take too long, or store some key objects in an array and check that they don't take too much space.

In this chapter, we'll first look at why it is possible to get away with a small number of simple load tests for a system like LabWizard. Then we will examine the very basic tools we have used in writing these tests: for measuring time, we present a simple Stopwatch class, and for measuring RAM usage we have a class called Size. As examples of load tests, we will use the single load test for GrandTestAuto and also a user interface load test for LabWizard. We finish with a short discussion of profilers and other tools.

What to Test

Writing load tests can be a lot of fun.

Typically, we write a test that exercises some part of our system and then run the test in a profiler. This reveals a few inefficiencies that can be ironed out. After re-working the relevant sections of code, we re-run the test and congratulate ourselves on the performance improvement.

This process has qualities that make it potentially addictive: there is a well-defined problem, a clear path to finding the solution and an obvious criterion for success.

We must take care not to get carried away.

 Extreme Testing Guideline: Only write load tests for the things that actually matter in the application.

Whether a performance issue actually matters depends on the specific (and perhaps contractual) requirements of our software, the severity of the problem and how often the problem occurs.

Once we have identified a performance issue that must be fixed, there remains the question "How much performance gain is needed?" Being perfectionists, we software engineers have a natural tendency to squeeze as much performance as possible out of the system just because we know it can be done. This process however is a prime example of the 'law of diminishing returns'—after the low hanging fruit, further performance gains take exponentially more effort to implement.

So again, let's not get carried away!

 Extreme Testing Guideline: Set load test pass / fail criteria at the minimum performance level required.

Before we look at techniques for load-testing our software, let's look at some real-life examples of performance problems in a deployed system, namely LabWizard. What we will see is that sometimes it just does not pay to fix things, even if we'd like to. And if we are not going to fix a performance problem, there is no point in writing a load test that quantifies it.

Overnight 'Housekeeping' Takes Too Long

The LabWizard server software is deployed on a dedicated machine, and has a 'housekeeping' routine that starts in the dead of night. This housekeeping involves database backups, cleanup and archiving of the day's transactions stored as files, sending logs to our support software center, and various other jobs that are best done at a quiet time. Until recently, we could afford this process to be inefficient, as no one was awake to notice that the CPU was maxed-out, and no cases were being interpreted for an hour or more. However, when we installed our software at a certain hospital, there was no time of the night where LabWizard could be offline for long and we had no choice but to attend to this issue. Fixing the problem turned out not to involve more efficient algorithms in terms of fewer CPU cycles. Rather, we had the same amount of work done by more threads. So, in this case, the nature of our server installations (dedicated multi-CPU machines) and the fact that the work done could be broken down into independent tasks meant that relatively simple code changes solved this performance issue.

Deleting Cases Takes Too Long

The LabWizard KnowledgeBuilder application contains several lists of cases. The following figure shows the Knowledge Builder's **Rejected** cases list:

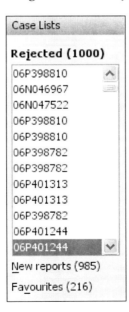

In an old version of LabWizard, when the user deleted lots of cases on a very large case list, it took several minutes for the hour glass to vanish and control to be returned to that user. This was a performance bug that we judged needed to be fixed straight away, as it made an important aspect of the software unusable. As with the housekeeping bug, this problem was fixed by running more server-side threads. The user action of deleting cases was changed to a user action that started a thread that deleted the cases in the background. This freed the GUI to handle other events in the meantime. We will look at the load test for this bug later in this chapter.

The BMD Server is Too Slow to Start

One of the LabWizard products is a reporting system for Bone Mineral Densitometry, which is the use of special X-ray machines to diagnose osteoporosis and other bone diseases. This system includes a server program that contains a patient database. The server communicates by RMI to various client applications and DEXA (Dual Energy X-ray Absorptiometry) scanners. If there are many thousands of patients, the server program can take up to ten minutes to start, during which time all the patient records are loaded into memory. This is an annoying delay during our testing and when we want to give demonstrations. In the 'real world', however, it is just not an issue as the server is running almost continually, and shutdowns can be scheduled for the convenience of the customer. So this is a situation where we've chosen to ignore (at least for now) a performance issue so we can fix other problems that really do matter.

Measuring Time

An issue with load testing is that this year's supercomputer is next year's clunker. If our test has an assertion that an operation should take less than, say, 20 seconds on our reference platform, do we need to revise our test when we run it on a faster machine?

If we answer 'yes' to this question, then we can't specify our load test as an absolute time, but as a relative time. This situation may arise, for example, if we are developing code for a highly constrained computing platform such as a traffic signals processor deployed on the curbside. We may need to ensure that any timings measured on our (generally higher performance) test machines can be extrapolated back to realistic timings on our deployed platform.

If we want or need to write our load tests in terms of relative time, we will have to concoct some standard task, measure how long it takes to complete on a particular machine, and then express our test assertions in terms of that time unit. For example, we might have as a standard task the parsing of a large XML document, the compression and decompression of a large file, the implementation of a communications protocol, or some other intensive operation to try and assess the performance difference between the test machine and the deployed platform. Better still would be to use a combination of such operations. Our load test then might assert that the operation under test must take less than say 100 of these relative time units.

This seems right, but is extremely difficult to do properly because the increase in performance of a new machine over an older one might be much greater in some areas than in others. Once we start down this path, we can easily spend more effort on designing our standard tasks than we do on writing load tests themselves.

Let's look back to our guideline that 'we only load test the things that matter'. If we have a test specified by a relative time of say 10 units, and if our new machine nominally has twice the performance as the old one, we need to ask ourselves whether it really matters if the time taken by the test is not quite down to 5 units.

Another important aspect of timing tests is the issue of scalability. If we do have an absolute time limit of say 10 milliseconds to do 10 operations, can we do 20 operations in roughly 20 milliseconds, and so on? In this case, our load test can be specified in terms of multiples of an absolute time limit.

So unless we're in the realm of hard real-time programming or developing for embedded processors, which is beyond the scope of this book, it's almost certainly sufficient to put absolute times, or multiples of some absolute time, in our load tests. In other words, load testing is not rocket science, unless we're doing rocket science.

Given this simplification, load tests really are simple to write, and rely on nothing more complex than `System.currentTimeMillis()`. One tool to make them even easier is the `Stopwatch` class, which we will now demonstrate.

GrandTestAuto (GTA), our preferred testing framework, and the subject of the next chapter, has a load test that uses `Stopwatch`. The specification of the test is:

> Create various package structures containing a mix of production, unit test, function test, and load test classes, such that all tests just return 'true' and such that the overall testing result is 'true'. Have these package structures of various sizes and levels of sub-packages, such that there are packages of about 600, 1200, 3000, and 6000 classes in all. Take separate measurements of the times to create and run GTA on these package structures. Check that the maximum creation time is less than 20 seconds and that the maximum run time is less than 30 seconds. Check also that the time taken to create GTA and to run the tests grows approximately linearly.

Load Testing

The structure of the test is:

- Create a timer for measuring the time to initialize the GTAs.
- Create a timer for measuring the time to run each GTA.
- For each of the configured packages to be tested, time how long it takes to make a GTA, then time how long it takes to run it.
- Make the assertions about the results.

Here is the test itself, with the use of `Stopwatch` highlighted:

```
/**
 * See the GrandTestAuto test specification.
 */
public class Load extends FTBase {
    private Stopwatch creationTimer;
    private Stopwatch runTimer;
    public boolean runTest() throws IOException {
        creationTimer = new Stopwatch();
        runTimer = new Stopwatch();
        //a51:   27 classes in all.
        runForConfiguredPackages( Grandtestauto.test51_zip );
        //a52: 85
        runForConfiguredPackages( Grandtestauto.test52_zip );
        //a60 169
        runForConfiguredPackages( Grandtestauto.test60_zip );
        //a53: 175
        runForConfiguredPackages( Grandtestauto.test53_zip );
        //a54 297
        runForConfiguredPackages( Grandtestauto.test54_zip );
        //a55 451
        runForConfiguredPackages(  Grandtestauto.test55_zip );
        //a61 651
        runForConfiguredPackages(  Grandtestauto.test61_zip );
        //a62 1653
        runForConfiguredPackages(  Grandtestauto.test62_zip );
        //a63 3367
        runForConfiguredPackages(  Grandtestauto.test63_zip );
        //a64 5985
        runForConfiguredPackages(  Grandtestauto.test64_zip );
        //Check that the creation times are not growing too fast.
        long ct6 = creationTimer.times().get( 6 );
        long ct7 = creationTimer.times().get( 7 );
        long ct8 = creationTimer.times().get( 8 );
```

```
            long ct9 = creationTimer.times().get( 9 );
            assert ct7 < 2.5 * ct6;
            assert ct8 < 2.1 * ct7;
            assert ct9 < 2 * ct8;
            //Check that the run times are not growing too fast.
            long rt6 = runTimer.times().get( 6 );
            long rt7 = runTimer.times().get( 7 );
            long rt8 = runTimer.times().get( 8 );
            long rt9 = runTimer.times().get( 9 );
            assert rt7 < 2.5 * rt6;
            assert rt8 < 2.1 * rt7;
            assert rt9 < 2* rt8;
            //Check that the absolute times are not too great.
            assert creationTimer.max() < 20000;
            assert runTimer.max() < 30000;
            return true;
        }
        /**
         * Expands the given zip of packages and creates and
         * runs a GTA for that package hierarchy, measuring
         * the creation and run times.
         */
        private void runForConfiguredPackages( String packageZip )
                throws IOException {
            File zip = new File( packageZip );
            String fileName = Helpers.expandZipAndWriteSettingsFile(
                    zip, true, true, true, null, false, true, null );
            creationTimer.start();
            Settings settings = new Settings( fileName );
            GrandTestAuto gta = new GrandTestAuto( settings );
            creationTimer.stop();
            runTimer.start();
            boolean result = gta.runAllTests();
            runTimer.stop();
            //Tests should all have passed.
            assert result;
        }
    }
```

The source code for Stopwatch and its test are in the packages jet.testtools and jet.testtools.test.

Measuring RAM Usage

Excessive memory use by a program can have many adverse effects. As the amount of memory used goes up, the JVM in which our program is executing will spend more and more time managing memory and less time actually executing our program. If the entire physical RAM is used, the system will become infuriatingly slow as the operating system starts paging. At worst, the program will run out of memory and crash.

If our system uses RMI, then the memory size of serialized objects can have a direct effect on the user-responsiveness of the system. An example of this comes from the LabWizard system. Recall the earlier example of lists of cases, for which deleting cases took an inordinately long time. Each of these cases is part of a patient's medical record. As the user scrolls up and down the case list, these medical records need to be retrieved from the server and displayed. When the connection between the client and server is slow, the user-responsiveness of the system depends on keeping the size of the serialized cases down.

Measuring the size of serializable objects can be done using the serialization mechanism itself. We have provided a utility class, called jet.testtools.Size, to do this. Size has a static method, sizeInBytes(), that works as follows:

```
public static int sizeInBytes( Serializable s) {
    try {
        ByteArrayOutputStream baos = new ByteArrayOutputStream();
        ObjectOutputStream oos = new ObjectOutputStream( baos );
        oos.writeObject( s );
        oos.flush();
        return baos.toByteArray().length;
    } catch (IOException e) {
        e.printStackTrace( );
        assert false : "Could not serialise object, as shown.";
    }
    return -1;
}
```

Once we know the maximum expected size for a serializable object, our load test can make the simple assertion:

```
int measuredSize = Size.sizeInBytes( objectUnderTest );
assert measuredSize < MAXIMUM_ALLOWED_SIZE;
```

Whilst it is difficult to get an accurate measurement of the memory our application is using (due to the effect of the garbage collector), it may be important to have a load test that at least sets some reasonable upper bound for a specified scenario. After all, we need to be confident that our application will run on a platform with a certain amount of physical memory.

Similarly, it is important to test for "memory leaks" for operations that are repeated many times and which, if not fixed, would eventually crash the application.

To assist with these measurements, `Size` has another static method, `getUsedMemory()`, implemented as follows (see "*Java Platform Performance: Strategies and Tactics*", Wilson and Kesselman):

```
public static long getUsedMemory() {
    gc(); //hint for the garbage collector to do its thing...
    gc();
    long totalMemory = Runtime.getRuntime().totalMemory();
    long freeMemory = Runtime.getRuntime().freeMemory();
    return totalMemory - freeMemory;
}
```

For example, once we know the maximum expected used memory for an operation, our load test can make the simple assertion:

```
for ( int i=0; i < 1000, i++ ) {
    //perform the operation
    operationUnderTest();
    //check whether used memory is in fact constant
    assert Size.getUsedMemory() < MAXIMUM_EXPECTED_MEMORY;
}
```

With the rapidly increasing capacity of disk drives or other storage devices, the size of our persisted objects (that is, database or other files) may not matter. If it does matter however, then our load test can make an assertion on the file size found using the `File.length()` method.

The Load Tests for LabWizard

A software package such as GrandTestAuto is so simple that load tests use only the very basic tools we have described above. Surprisingly, these are precisely the same tools that we use for load testing LabWizard. This is because of the nature of the software and because a lot of the code that we use in unit and function tests can also be used in load tests.

For example, here is a LabWizard load test that checks that deleting cases from a list of more than 20,000 cases is responsive:

```
public class LotsOfCases extends FunctionTest {
    private String project = "ManyCases";
    public boolean runTest () throws Exception {
        //Start Knowledge Builder client,
```

```
            //open the configured KnowledgeBase
            //and go to the archive list.
            TestBase.setupConfiguredKB( "20301Cases", project );
            KBTestProxy kb = TestBase.startKBClient();
            kb.project().open( project );
            kb.focusToArchiveList();
            UICaseList archiveList = kb.archiveCaseList();
            //Go to the last case on the list.
            kb.uiRobot().end();

            //Check that deleting cases is responsive.
            kb.uiRobot().shiftPageUp( 10 );
            Stopwatch timer = new Stopwatch();
            timer.start();
            archiveList.delete();
            timer.stop();
            Long timeTaken = timer.times().get( 0 );
            assert timeTaken < 5000 : "Took: " + timeTaken;
            return true;
        }
    }
```

Because it makes use of the UI Wrapper classes that we developed for our unit tests, the code for this test is short and simple. Server-side load tests tend to be just as simple, merely starting a timer, calling a method, stopping the timer, and checking the elapsed time.

Profilers and Other Tools

If we know that some aspect of our software is too slow or takes too long and this needs to be fixed, our best first step is to write a test that replicates the problem. By running this test in a profiler, we can work out exactly what is causing the problem, and fix it. (Of course, the test should stay in our test suite even after fixing the problem, as insurance against it re-occurring.)

Sometimes it is not possible to run our software in a profiler. This is usually due to the fact that the software itself is consuming so much in the way of system resources that the profiling overhead brings it to a grinding halt. In a situation such as this, one option is to log the critical steps in our software's operation. By looking carefully at the log files, we might be able to find a small part of the code that is inefficient and which can be run in a profiler.

With GrandTestAuto and LabWizard, we could write load tests very easily because the things that needed to be measured were method invocations or user actions, both of which are readily performed in tests. If we are running a web application, however, our software is running inside a web server and there is no user interface to be activated. For load testing this kind of software, we can use Apache JMeter (see `http://jakarta.apache.org/jmeter/`) or something similar.

Summary

We should only write load tests for performance issues that really affect our end users. When we do write such a test, we can often do no better than measuring the time taken or the memory used by our software in performing the problematic task, and then simply asserting that the measured value is less than some known acceptable limit. For measuring time we can use the `Stopwatch` class, and for measuring memory usage we can use the `Size` class. These are both found in the package `jet.testtools`.

19
GrandTestAuto

One of the key packages in the LabWizard codebase consists of almost three hundred classes that define the operators in the LabWizard condition language. When we were developing this package we were faced with the problem of making sure that every single method was tested. This was before we had as extreme an attitude about testing as we do now, but we were more careful about this package than others because the methods directly related to the evaluation of the conditions in rules. To deal with this situation, we created a tool that not only ran the unit tests but at the same time checked that each public or protected method had a test method. This tool evolved into GrandTestAuto.

In this chapter we explain how software can be tested with GrandTestAuto, and the advantages gained by using the tool.

The complete requirements and design specifications, source code, test code, and test data for GrandTestAuto are supplied with the software for this book and can also be downloaded from the website http://www.grandtestauto.org.

What is GrandTestAuto?

GrandTestAuto (GTA) is a command-line tool for running the unit, function, and load tests in a Java code base. GTA uses the Java Reflection API and simple naming patterns to locate and run tests, either in a single JVM or across several JVMs on one or more machines on a network.

Let's look at the structure of a code base that is being tested using GTA. Here is a truncated and simplified version of the code base for the LabWizard product suite:

Package	Class
`rdr.attribute`	`Attribute (abstract)`
	`DerivedAttribute`
	`PrimaryAttribute`
`rdr.attribute.test`	`UnitTester`
	`DerivedAttributeTest`
	`PrimaryAttributeTest`
`rdr.bmd`	`Patient`
	`Scanner`
	`PatientManager`
`rdr.bmd.test`	`UnitTester`
	`PatientTest`
	`ScannerTest`
	`PatientManagertest`
`rdr.bmd.functiontest`	`RegisterNewPatient`
	`RegisterExistingPatient`
`rdr.bmd.loadtest`	`TimeToStartPatientManager`
	`PatientSerialisationSize`

The package `rdr.attribute` defines three classes, `Attribute`, which is abstract, and two concrete subclasses. The unit tests for these classes are in a package called `rdr.attribute.test`. This package contains a `UnitTester` class as well as tests for the two concrete classes.

The package `rdr.bmd` is more extensive. The reason for this is that it implements a major application (the Bone Mineral Density Reporting System) within the LabWizard product suite. This application is defined by a BMD Requirements Specification and associated BMD Test Specification. The function and load tests defined by the BMD Test Specification are implemented in the additional two test sub-packages, `functiontest` and `loadtest` respectively. The unit tests for the BMD classes are in the `test` sub-package. In addition, the `test` sub-package contains the BMD-specific UI proxy classes which are used at all three levels of BMD testing.

GTA is invoked by a call to its main class with the name of a parameters file as the only argument:

```
org.grandtestauto.GrandTestAuto GTASettings.txt
```

The parameters file is just a properties file with about a dozen key-value pairs; all but one are optional. The required parameter is the location of the classes directory for the project under test. So here is a bare-minimum settings file for GTA:

```
CLASSES_ROOT=..\\classes
```

GTA searches the classes directory for test packages. The naming pattern for these packages is simple: unit tests packages have name ending ".test", function test packages have name ending ".functiontest", and load test packages have name ending ".loadtest".

Within each unit test package, GTA will seek a class called `UnitTester` that implements `org.grandtestauto.UnitTesterIF`:

```
/**
 * All <code>UnitTester</code>s must implement this interface.
 * Additionally, they must have a public constructor from a
 * <code>GrandTestAuto</code> or a no-args constructor.
 */
public interface UnitTesterIF {
    /**
     * Runs the unit tests for the containing package.
     *
     * @return <code>true</code> if and only if all
     * the tests passed.
     */
    public boolean runTests();
}
```

GTA uses Reflection to instantiate these classes, then executes `runTests()` and records the result. The overall unit testing result is the conjunction (logical AND) of the individual package results. If any package is without a unit-test package, the overall result is false.

After the unit tests have been run, the function tests are invoked and run. As discussed in Chapter 17, each function test is a single class that tests our software against a single requirement. In order to be detected and run by GTA, these implement `org.grandtestauto.AutoLoadTest`:

```
/**
 * Function and Load tests should implement this interface
 * and be in a package with name ending ".functiontest" or
 * ".loadtest". Additionally, they should have a public
 * no-args constructor.
 */
public interface AutoLoadTest {
    public boolean runTest() throws Exception;
}
```

For each function test package, GTA finds all the classes implementing `AutoLoadTest`, instantiates them, and then invokes `runTest()`. The result for the package is the conjunction of the results of the tests within it, and the overall function testing result is the conjunction of the results for the individual function test packages.

Finally, the load tests are run. Like the function tests, these implement the interface `org.grandtestauto.AutoLoadTest` and the running and results calculation for load tests is the same as for function tests.

We run unit tests before function tests because if the unit tests do not pass, the function tests are unlikely to either. The load tests are run after the unit and function tests because correctness comes before performance. Generally, unit tests run very quickly, function tests are somewhat slower, and load tests can take a very long time indeed. In an automated build environment, GTA is set to 'fail fast' so that no time is wasted running tests on a code base that is already known to have faults. The overall testing result is the conjunction of the unit, function, and load test results. In summary:

- Tests are run automatically and there are no hard-coded lists of tests.
- The three levels of tests are run in a natural order of unit, function, and load tests.
- There are few constraints on the tests themselves.
- The overall test result will not be 'true' if there are any packages that are not unit-tested.

Unit Test Coverage

By using GTA, we are at least ensuring that each package in our code base has a unit test package. However, GTA goes further than this — GTA will ensure that:

- Every public non-abstract class has a unit test.
- All of the public and protected methods are tested.
- There are tests for all public constructors.

There are some exceptions to these rules, for example `Enums` do not generally need a test, and GTA cannot test protected methods that are final (so these should be avoided if possible). The rules themselves are stated more carefully in the GTA Requirements Specification. What is important is that they encode the 'test everything' policy that we worked out in Chapter 1.

Advantages of Using GTA

Why try to improve on other testing frameworks, such as JUnit? GTA has several advantages, which we'll explore in this chapter. Briefly though, some of the main reasons for using GTA are as follows.

Firstly, GTA can be used to ensure complete unit-test coverage of our code base. The JUnit view is that we should "just test everything that could reasonably break" (according to the JUnit FAQ Best Practices section). Our experience as developers is that this is everything.

Secondly, GTA detects and runs tests automatically. Some testing frameworks, such as older releases of JUnit, rely on explicit listing of the test classes to be run as a suite. Our experience is that in systems with hundreds or thousands of test classes, some classes will not be listed. In fact, an early version of GTA had this deficiency. When we converted to the new version, we found a disturbing number of tests that were not being run.

Thirdly, the runtime parameters for GTA, which are detailed in this chapter, provide an enormous amount of flexibility over which tests are run. For continuous integration testing and final release testing, we will of course want to run all tests. But sometimes we just want to concentrate our testing on particular packages, or a particular testing phase (unit, function, or load). The next chapter gives more examples of situations where we just want to test particular packages, classes, or methods within a class.

A fourth advantage of GTA is that it can be used to automate the distribution of tests over a number of machines. Using this, we can make the most of all the test machines available and run the tests as quickly as possible. We can also use this system to allocate certain test packages to particular machines, if this is needed. We will explore distributed testing later in this chapter.

Finally, GTA is extremely simple to use. The tool really does nothing more than search for test classes and run them using Reflection. At first glance, it might seem that this limits the expressiveness of the tests. In fact, this is not the case at all, rather it means that any setup before tests are run, and cleanup after the tests, is expressed in plain Java code rather than by using XML files or special annotations (the tool TestNG is particularly guilty here). The design philosophy with GTA is that tests are to be written by software engineers in Java, and there is nothing more expressive for organizing the tests than Java itself.

Getting Started

Let's have a look at how GTA's unit test coverage works in practice.

Suppose that we are developing a package `pack`. The first class to be written is `A`:

```
package pack;
public class A {
    public A( String str ) {
        //Constructor body ...
    }
    public String a() {
        //Method body ...
    }
    public String b( String x ) {
        //Method body ...
    }
}
```

The unit test package for `pack` is `pack.test`, and into it we add a `UnitTester` class that extends `CoverageUnitTester`:

```
package pack.test;
import org.grandtestauto.*;
public class UnitTester extends CoverageUnitTester {
    public UnitTester( GrandTestAuto gta ) {
        super( gta );
    }
}
```

`CoverageUnitTester` is a class that comes with GTA, and as its name implies, does the work of enforcing the unit-test coverage policy developed in Chapter 1, by which every accessible method requires an explicit test. There are rare situations when our `UnitTester` class does not extend `CoverageUnitTester`. Such situations generally arise when we are integrating legacy packages.

The unit test for `A` must be in the package `pack.test` and must be called `ATest`:

```
package pack.test;
public class ATest {
    //Class body..
}
```

We need to have a test method for each of the methods in A and also for the constructor. The test for a method `m()` is called `mTest` and is public, takes no arguments, and returns a `boolean`. (We will come back to test method naming later, to discuss how to deal with multiple methods with the same name and testing classes with more than one constructor.) The constructor test is like method tests in that it is public, has no parameters, and returns a `boolean`. It is called `constructorTest`.

```
package pack.test;
public class ATest {
    public boolean aTest() {
        //Test method body ...
    }
    public boolean bTest() {
        //Test method body ...
    }
    public boolean constructorTest() {
        //Test method body ...
    }
}
```

At this point we can run our tests. We create a settings file that tells GTA where the classes are and how we want the test results recorded. At this early stage, we don't want to log our results in a file, just to the console:

```
CLASSES_ROOT=..\\classes
LOG_TO_CONSOLE=true
LOG_TO_FILE=false
```

If this file is called `GTASettings.txt`, then the command to invoke GTA is:

```
java -cp ..\classes;ExtremeTesting.jar org.grandtestauto.GrandTestAuto
                                                         GTASettings.txt
```

Assuming that the tests passed, the output would be something like:

```
GrandTestAuto 3.1
########## Running Unit Tests ##########
ATest
    aTest passed
    bTest passed
    constructorTest passed
>>>> Results of Unit Tests for example1: passed. <<<<
******* Overall Unit Test Result: passed. *******
```

Suppose that we add a new public method `c()` to A. If we run our tests again, the result will be `false`, and a message that `A.c()` is untested will be printed. Adding a method `cTest()` to the class `ATest` fixes this.

Now suppose that we create a new class B:

```
package pack;
public class B extends A {
    public B( String str ) {
        //Constructor body ...
    }
    public String a() {
        //Method body ...
    }
    public String d( String x ) {
        //Method body ...
    }
}
```

If we invoke GTA at this point, our tests will fail because there is no test class for B. The unit test for B must test precisely the methods of B that are defined in the package of B and are not already tested. That is, we must test the constructor, the overridden version of `a()` and also the new method `d()`. We do not need to test the methods inherited from A, as these are tested in `ATest`, or those inherited from `java.lang.Object`, which we assume have been tested by Sun. So here is a minimal `BTest`:

```
package pack.test;
public class BTest {
    public boolean aTest() {
        //Test method body ...
    }
    public boolean dTest() {
        //Test method body ...
    }
    public boolean constructorTest() {
        //Test method body ...
    }
}
```

As the `UnitTester` runs, it records the names of the test methods it executes. When all of the tests are complete, a check is done to match the test methods against the methods declared in the classes in the package being tested.

In summary:

- We must provide a test class for every non-abstract public class.
- We must provide a test method for every accessible method.
- All the test methods in all the test classes get run automatically.
- We don't need to write tests for methods that are inherited from outside the package we're testing.

Testing Overloaded Methods

Now consider a class with several methods having the same name:

```
package example3;
public class C {
    public C(String s) {
        //Constructor body ...
    }
    public String m() {
        //Method body ...
        return null;
    }
    public String m( String x ) {
        //Method body ...
        return null;
    }
    public String m( int n ) {
        //Method body ...
        return null;
    }
    public String m( int[] x ) {
        //Method body ...
        return null;
    }
    public String m( int[][] x ) {
        //Method body ...
        return null;
    }
}
```

We want to have a test for each version of `m()`. How is this achieved? The methods differ in their parameter lists, so we include some details of the parameters in the names for the test methods:

```
package example3.test;
public class CTest {
    public boolean constructorTest() {
        //Method body ...
        return true;
    }
    public boolean mTest() {
        //Method body ...
        return true;
    }
    public boolean m_String_Test() {
        //Method body ...
        return true;
    }
    public boolean m_int_Test() {
        //Method body ...
        return true;
    }
    public boolean m_intArray_Test() {
        //Method body ...
        return true;
    }
    public boolean m_intArray2_Test() {
        //Method body ...
        return true;
    }
}
```

The exact rules for test method name-mangling are explained in the GTA Requirements Specification.

The mangled names are pretty ugly, but remember that name mangling is not needed if there is only one version of a method that is to be tested. We do get horrible test methods such as

```
constructor_ObjectId_String_SampleSequenceCase_Test
```

However, these are exceptions to the norm. In fact, our experience of writing GTA tests for previously untested packages is that if there are a lot of methods with the same name, often one or more of them will turn out to be 'helpful' variants that are never used and should be deleted. That is, a little bit of code pruning can mean that name-mangling is not needed. To be really convinced that name-mangling is not unmanageable, we recommend a look at the tests in the code accompanying this book. Another thing to bear in mind is that we've unit-tested a codebase containing well over a thousand classes without it annoying us too much.

The name-mangling scheme will not work on classes where there are variants of a method that cannot be distinguished by the unqualified names of the parameters, for example:

```
public String m( org.foo.String x ) {
    //Method body ...
}
public String m( String s ) {
    //Method body ...
}
```

We have only once seen this happen, and the problem was resolved by renaming one of the methods, which made the code more readable anyway.

Testing Protected Methods

Suppose that a class D has a protected method:

```
package example3;
public class D {
    protected void p() {
        //Method body ...
    }
}
```

How can this be tested? The trick is to define an extension of D that exposes the method to the test package:

```
package example3.test;
import example3.*;
public class DTest {
    public boolean constructorTest() {
        return true;
    }
    public boolean pTest() {
```

```
        DExt d = new DExt();
        String val = d.p();
        //Rest of test...
        return true;
    }
    private class DExt extends D {
        public String p() {
            return super.p();
        }
    }
}
```

Of course, if the protected method is final, this trick does not work, which is why GTA does not insist that final protected methods be tested.

Extra Tests

We often want to write test methods that do not correspond precisely to a single method or constructor in the class under test. For example, we may need a test that involves several methods from the class. These tests just need to follow the naming and signature convention (no args, return type `boolean`, and the method name to end in "Test"), and they will be run automatically.

At a larger scale, we sometimes want to add extra unit test classes to a test package. Again, these tests are often "integration-type" tests which involve several classes from the package. As long as the test class has a name ending with "Test" and a no-args constructor, it will be created and run automatically.

Classes That Do Not Need Tests

If we are using generated code, then we will probably not want explicit unit tests for the generated classes, though of course the code generation mechanism must be tested. There is a simple annotation that can be used to tell GTA that a class is excused from explicit testing:

```
package org.grandtestauto;
import java.lang.annotation.*;
/**
 * This annotation is for marking classes that do not
 * need to be tested.
 */
@Retention(RetentionPolicy.RUNTIME)
public @interface DoesNotNeedTest {
    String reason();
}
```

Day-To-Day Use of GrandTestAuto

Whatever our testing platform is, we need to apply it in different ways: as part of our automated build process, when working on a particular class, when verifying a package before checking it in to source control, and so on. Let's look at how GrandTestAuto supports these kinds of operations.

Running Just One Level of Test

Suppose that we only want to run unit tests, or just function and load tests. We can control the phases of tests that are run by setting the properties

```
RUN_UNIT_TESTS
RUN_FUNCTION_TESTS
RUN_LOAD_TESTS
```

The default values for these are `true`.

Running the Unit Tests for a Single Package

Suppose we want to run all unit tests for a specific package before checking it in. To do this, we use the

```
SINGLE_PACKAGE
```

parameter. For example, to run just the package `rippledown.attribute.command.test`, we would use these settings:

```
CLASSES_ROOT=..\\classes
SINGLE_PACKAGE=rippledown.attribute.command
RUN_UNIT_TESTS=true
RUN_FUNCTION_TESTS=false
RUN_LOAD_TESTS=false
```

GTA allows for a sort of package name completion (more on this later), and `booleans` can be abbreviated with `'t'` and `'f'`, so we can write these settings more succinctly:

```
CLASSES_ROOT=..\\classes
SINGLE_PACKAGE=r.a.c
RUN_UNIT_TESTS=t
RUN_FUNCTION_TESTS=f
RUN_LOAD_TESTS=f
```

Running the Unit Tests for a Single Class

If `SINGLE_PACKAGE` is set then we can also set a value for

> `SINGLE_CLASS`

For example, if we just wanted to run the unit test for `rippledown.attribute.command.test.AttributeMove` we would use these settings:

> `CLASSES_ROOT=..\\classes`
> `LOG_TO_FILE=false`
> `RUN_UNIT_TESTS=true`
> `SINGLE_PACKAGE=r.a.c`
> `SINGLE_CLASS=AttributeMove`

Running the Tests for a Selection of Packages

By setting the property,

> `FIRST_PACKAGE`

we restrict testing to the specified package, plus those that come after it in alphabetical order. Likewise,

> `LAST_PACKAGE`

restricts testing to those packages up to and including the specified package. If both of these properties are set, just the tests between and including the named packages will be run. These properties are over-ridden by `SINGLE_PACKAGE`.

Something to note is that these three properties, as far as unit testing go, refer to packages to be tested, whereas for function and load tests, they refer to packages of tests themselves. This means that if we set

> `FIRST_PACKAGE=com.aardvark`

and

> `LAST_PACKAGE=com.aardwolf`

then these packages will be run:

> `com.aardvark.test`
> `com.aardvark.functiontest`
> `com.aardvark.loadtest`
> `com.aardwolf.test`

If we want to run just the `aardvark` packages, we set

    LAST_PACKAGE=com.aardvark.loadtest

and if we want to include the `aardwolf` function and load tests, we set

    LAST_PACKAGE=com.aardwolf.loadtest

Package Name Abbreviation

The values for `FIRST_PACKAGE`, `LAST_PACKAGE`, `SINGLE_PACKAGE` can use a simple abbreviation scheme as follows. For example, we can abbreviate the package

    org.grandtestauto.test.functiontest

to

    o.g.t.f

If more than one package name matches the abbreviation, the first one in alphabetical order will be chosen.

Running Tests for a Selection of Classes Within a Single Package

The parameters

    FIRST_CLASS
    LAST_CLASS
    SINGLE_CLASS

are used to run a range of tests, or just a single test, within a single package. They only have an effect if `SINGLE_PACKAGE` has been set. As before, for unit tests the values of these parameters refer to classes to be tested. However, for function and load tests the parameters refer to the test classes themselves.

Running Individual Test Methods

If the `SINGLE_PACKAGE` and `SINGLE_CLASS` properties have been set, then we can specify which test methods are to be run within the test class. The relevant parameters are

    FIRST_METHOD
    LAST_METHOD
    SINGLE_METHOD

As expected, 'First' and 'Last' refer to alphabetical ordering of the test methods, and the 'Single' parameter overrides the other two. These parameters are only applicable for unit tests.

Running GTA From Ant or CruiseControl

Apache Ant (see http://ant.apache.org) is the most popular Java build tool, and GTA comes with an Ant task that we can invoke to run our tests. For example, here is an extract from the build file for this book's code:

```
<target name="runtests" description="Run GrandTestAuto">
    <taskdef name="run-gta"
        classname="org.grandtestauto.ant.RunGrandTestAuto">
        <classpath>
            <pathelement path="${classes}"/>
            <fileset dir="lib">
                <include name="*.jar"/>
            </fileset>
            <pathelement path="./testtemp/"/>
        </classpath>
    </taskdef>
    <run-gta SettingsFileName="GTASettings.txt"/>
</target>
```

The continuous build automation tool, CruiseControl (see http://cruisecontrol.sourceforge.net) uses Ant, so GTA can be run from it too.

GTA Parameters

Here is a complete list of the settings file properties accepted by GTA.

Key	Explanation
CLASSES_ROOT	The root directory for the classes to be tested. This value is required.
	On Windows, express file paths using '\\\\' or '/' for '\\'.
FAIL_FAST	If true, testing is stopped as soon as possible after the first test failure. Default is false.
LOG_TO_FILE	Should the results be logged to file? Default is true.
LOG_FILE_NAME	The file in which results are written, if logging to file. Default is 'GTAResults.txt'.
LOG_TO_CONSOLE	Should the results be logged to the console? Default is true.
RUN_UNIT_TESTS	Are unit tests to be run? Default is true.

Key	Explanation
RUN_FUNCTION_TESTS	Are function tests to be run? Default is true.
RUN_LOAD_TESTS	Are load tests to be run? Default is true.
FIRST_PACKAGE	The name of the package from which tests should be run (inclusive).
LAST_PACKAGE	The name of the package to which tests should be run (inclusive).
SINGLE_PACKAGE	For running just a single package of tests. Overrides FIRST_PACKAGE and LAST_PACKAGE.
FIRST_CLASS	Determines the first test to run, if SINGLE_PACKAGE is set. For unit tests, set this to the class to be tested, for function and load tests use the name of the test.
LAST_CLASS	Determines the last test to run, if SINGLE_PACKAGE is set.
SINGLE_CLASS	Determines the only test to run, if SINGLE_PACKAGE is set. Overrides FIRST_CLASS and LAST_CLASS.
FIRST_METHOD	The first test method to run, if SINGLE_PACKAGE and SINGLE_CLASS are set. Only relevant for unit tests.
LAST_METHOD	The last test method to run, if SINGLE_PACKAGE and SINGLE_CLASS are set. Only relevant for unit tests.
SINGLE_METHOD	The only test method to run, if SINGLE_PACKAGE and SINGLE_CLASS are set. Only relevant for unit tests.

Distributed Testing Using GTA

GrandTestAuto implements a distributed testing system in which tests can be run on several computers at once. This allows us to make the most of available test machines. Another advantage is that each test package is now run in a dedicated JVM, so the failure of one test package is less likely to have knock-on effects for subsequent test packages than in a system where tests are run in a single JVM.

How it Works—In Brief

A GTA distributed testing system consists of client programs together with a single server program, all running on various computers on a network. (It is also possible to have more than one client on a physical machine, so long as the tests do not interfere with one another.)

The computers on which the client programs run, and the server computer, all need to have access to a shared directory, which contains the classes to be tested.

The client programs are instances of org.grandtestauto.distributed.TestAgent. Each TestAgent has an internal thread which executes a repeated pattern of finding a server, retrieving a test task, executing that task in a separate JVM, and then reporting the results to the server.

The server program is an instance of org.grandtestauto.distributed.GTADistributor.

At startup, a GTADistributor works out which packages are to be tested. It then waits for connections from the TestAgents, assigns testing jobs to them, and receives the results of these tests. When all packages have been tested, the results are collated and printed out. The GTADistributor then shuts down.

In general, the TestAgent programs are continually running, and spend a lot of time looking for a server from which to request work. A GTADistributor program is brought into existence when there is a new software build to test. When the tests are finished, it shuts down. Thus the TestAgents outlive many GTADistributors.

The distributed testing system includes a notion of 'grade' that gives some control as to which tests are run by which clients. The essence of the grade system is that each package can be assigned a grade, and each TestAgent is assigned a maximum grade of test that it can run. When a client requests work, it is given the highest-graded job that it can manage. In this way, the long-running jobs are done first, which gives the highest test throughput. The grade system can also be used to give very fine-tuned control over the distribution of test classes and can ensure that particular packages are assigned to particular machines.

More details can be found in the user documentation for GTA.

A Distributed Testing Example

The code used in this book is tested with a distributed GTA system on a small network consisting of two machines, called 'wallaby' and 'numbat', The latter is the main development machine. The development directory on numbat is called **code** and has the following structure:

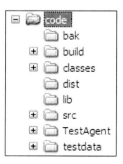

The **code** directory is shared on the network and is mapped as the 'T' drive on both `numbat` and `wallaby`. According to this mapping, the classes to be tested are in `T:\classes`. The `GTADistributor` is run on `numbat` from a batch file in the **build** directory. The batch file is:

```
set LIB=T:\lib
c:\jdk1.6.0\bin\java -ea -Xmx256M -cp T:\classes;T:\build\
testtemp;%LIB%\activation.jar;%LIB%\jh.jar;%LIB%\PDFBox.jar;%LIB%\
FontBox.jar;%LIB%\commons-io.jar;%LIB%\mail.jar;%LIB%\jxl.jar; org.
grandtestauto.distributed.GTADistributor DistributorSettings.txt
```

The settings file for the `GTADistributor` contains:

```
CLASSES_ROOT=..\\classes
DISTRIBUTOR_PORT=8086
```

We run `TestAgent`s on both `wallaby` and `numbat`. The settings file for the `TestAgent` on `wallaby` is:

```
BASE_DIR=.
CLASSES_ROOT=T:\classes
NAME=WallabyAgent
MAXIMUM_GRADE=10
SERVER_ADDRESS=numbat
SERVER_PORT=8086
JVM_OPTIONS_LIST=-ea -Xmx512M -DTestDataRoot="T:\testdata"
```

The last line defines the flags for the test JVMs and can be explained as follows. First of all, the flag `ea` is because we want assertions enabled. Next follows a memory setting. The third flag sets the value of the `TestDataRoot` system property: configured files used in tests are taken from this location. Thus, in this distributed testing setup, the configured test data and the classes being tested are taken from a central location.

Summary

GrandTestAuto is the ideal tool for implementing our 'test everything' policy for unit-testing. As well, it can run function tests and load tests and can be used to distribute the tests over a number of machines.

There are lots of examples of tests written for GTA, including those of GTA itself, in the source code.

20
Flaky Tests

Tests that fail intermittently are the biggest obstacle to an automated testing process. As we deal with them all the time, we even have a nickname for them. We call them 'flaky' tests.

The problem with having flaky tests is that they remove the element of reproducibility from our testing process. If our tests pass, we don't know whether all the bugs we're testing against have actually been fixed, or whether they just happened not to manifest themselves in that testing run. If our tests fail, this could be due to either production errors or testing errors, or both.

If we can't be sure of our test results, we can't be sure of our software.

Furthermore, even if we are fairly confident that the "flakes" are due to testing errors (for example a test that fails because a previous test did not clean up properly), the flaky tests are still a problem since they prevent the testing process from being fully automated.

In this chapter, we'll look at ways of tracking down and fixing flakes. As well, we'll investigate the converse problem of writing tests that reliably break software that itself intermittently fails. Our final topic will be how to identify and fix tests that do not terminate.

A Flaky 'Ikon Do It' Unit Test

Let's begin with a real example from the unit tests of 'Ikon Do It'.

The test method `IkonMakerTest.newIkonTest()` had occasional failures. The symptoms were:

- The test always passed when run by itself, even if it was run many times in a single JVM invocation.
- It always passed when all of the tests in `IkonMakerTest` were run.
- It would occasionally fail when all tests in its containing package were run.

Flaky Tests

This failure pattern was really annoying because it meant that it always took several minutes to reproduce the error. Unfortunately, once all of the more obvious problems have been dealt with, these kinds of flakes remain.

The test itself is as follows. An `IkonMaker` is started and a new 16-by-16 icon created:

```
init();
//First a simple one. Build it and save and check the image.
ui.createNewIkon( "ike1", 16, 16 );
```

The icon at this point consists of 256 pixels that are all the default background color. Four pixels are re-colored:

```
Color colour = new Color( 55, 66, 222 );
ui.selectColour( colour );
ui.uiIkonCanvas( 16, 16 ).clickPixelSquare( 5, 6 );
ui.uiIkonCanvas( 16, 16 ).clickPixelSquare( 6, 7 );
ui.uiIkonCanvas( 16, 16 ).clickPixelSquare( 7, 8 );
ui.uiIkonCanvas( 16, 16 ).clickPixelSquare( 8, 9 );
```

The icon is exported as a PNG image:

```
File dest = new File( Files.tempDir(), "ike1.png" );
ui.export( dest );
```

A two-dimensional array of colors representing the expected image is built:

```
Color[][] expected = new Color[16][16];
for (Color[] anExpected : expected) {
    for (int j = 0; j < anExpected.length; j++) {
        anExpected[j] = IkonMaker.DEFAULT_BACKGROUND_COLOUR;
    }
}
expected[6][5] = colour;
expected[7][6] = colour;
expected[8][7] = colour;
expected[9][8] = colour;
```

Finally, the exported PNG image is compared with the expected array:

```
displayAndCheckSavedImage( dest, expected );
```

This method first creates a frame safely (in the event thread):

```
final JFrame[] frames = new JFrame[1];
UI.runInEventThread( new Runnable() {
    public void run() {
        frames[0] = new JFrame();
    }
} );
```

Chapter 20

and then puts the image file into a `JLabel` in the frame:

```
final JLabel[] result = new JLabel[1];
UI.runInEventThread( new Runnable() {
    public void run() {
        result[0] = new JLabel();
        result[0].setIcon( new ImageIcon(
                Toolkit.getDefaultToolkit().getImage(
                        imageFile.getAbsolutePath() ) ) );
        frames[0].add( result[0] );
        frames[0].pack();
        frames[0].setVisible( true );
        frames[0].toFront();
    }
} );
```

`Cyborg` is then used to check each of the displayed pixels:

```
final JComponent imageComponent = result[0];
for (int i = 0; i < expectedPixels.length; i++) {
    Color[] expectedPixel = expectedPixels[i];
    for (int j = 0; j < expectedPixel.length; j++) {
        Color color = expectedPixel[j];
        cyborg.checkPixelInComponent(
                i, j, color, imageComponent );
    }
}
```

Finally, the frame is disposed safely:

```
//Close the frame showing the image.
UI.runInEventThread( new Runnable() {
    public void run() {
        frames[0].dispose();
    }
} );
```

If the test runs as expected, then just before the frame is disposed, there is a tiny frame showing the icon in the top left hand corner of the screen. Behind this frame, the IkonMaker also shows the icon, at its much larger drawing-mode size.

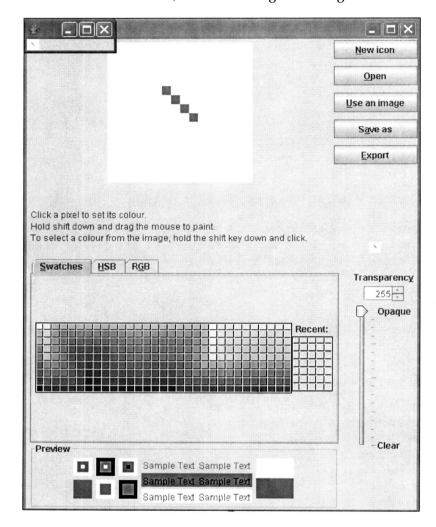

When the flake manifested itself, these screens did not appear, and the test failed with an error message saying that there was a pixel that did not have the expected color.

Our initial assumption was that the image in the JLabel was taking some time to load and that the display-and-check code was not thread-safe. As a workaround, we put in some code that waited for the expected pixels, and threw an assertion error if this timed out. This did not remove the flake, nor did several other similar changes to the test.

Chapter 20

What was needed was a more reliable and faster means of producing the problem. CruiseControl and GTA came to the rescue here. By experimenting with GTASettings.txt, we eventually discovered that running just IkonCanvasTest and IkonMakerTest would reproduce the problem fairly reliably. It would still take a few minutes each time to get the problem, but at least we were now in a position where we could start CruiseControl and then just wait for the error to show up.

The next step was to pause on failure so that we could inspect the state of the broken system. When we did this, it looked as though the wrong file was being displayed, though in fact the file name was correct. The following figure shows a test run that failed because of the flake.

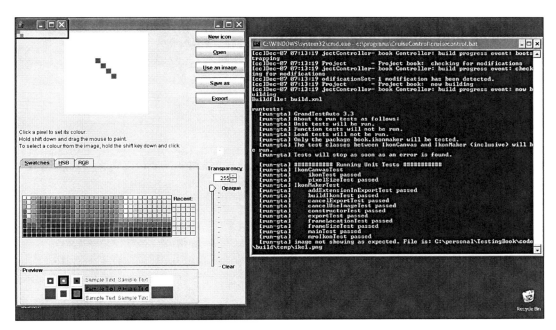

When we looked at the saved file using another application, the expected image was displayed. So there must have been some caching happening somewhere. Sure enough, we were creating the ImageIcon in the JButton using the wrong method call. Our code was:

```
result[0].setIcon( new ImageIcon(
      Toolkit.getDefaultToolkit().getImage(
            imageFile.getAbsolutePath() ) ) );
```

The `getImage()` method does the caching. A previous test had displayed another image file, and the same pixels were being displayed now. By changing the `getImage()` call to `createImage()`, the problem was solved. By the way, the Javadoc for `getImage()` is quite clear about the caching, we should have paid more attention to it.

Experiences like this can teach us a lot. We wasted a fair bit of time by hoping that the flake was caused by a threading problem, and trying various fixes, without really understanding what was going on. Making semi-random code changes, and hoping the problem will go away, is never going to work. It's better to work at understanding the root cause of the problem right from the start, even if this seems like a distraction from the 'real' job of getting software completed. (But of course our job is to produce a system for producing new versions of our software, not just a one-off deliverable, so effort on the automated building and testing of our software is well-spent.) Using CruiseControl and GTA to reliably reproduce the error, and pausing on failure, was an important strategy. We'll discuss the role of these tools later on.

Writing Reliable Tests

When we have an intermittent problem in our software, the challenge is to write a test that demonstrates it reliably. If we can achieve this, then when the test passes we can be reasonably sure that we've fixed the problem.

We recently had a bug like this in our LabWizard software. We had used code from a JavaWorld article (Java Tip 87: *"Automate the hourglass cursor"* By Kyle Davis, JavaWorld.com, 02/21/00) to implement a wait cursor. The code worked by replacing the default event queue with a custom queue that incorporated the wait cursor.

This worked most of the time, but occasionally the wait cursor would be showing inappropriately. Another article (The Java Specialists' Newsletter [Issue 075], 2003-07-29, *"An Automatic Wait Cursor: WaitCursorEventQueue"* by Nathan Arthur.) analyzed the JavaWorld code and fixed the problems. The new code had some unit tests, but these came with the warning that they would occasionally fail even though there were no bugs. This is exactly the sort of situation we want to avoid.

Instead, we wrote a test that always failed when using our old version of the code. Firstly, a `JFrame` containing a single button was shown:

The button was activated, which resulted in a `JDialog` showing an **OK** and **Cancel** button:

As soon as the button in the dialog got the focus, the *Escape* key was pressed. A check was then made for the kind of cursor showing in the frame. The code to get the cursor type was:

```
    private int frameCursorType() {
        final int[] resultHolder = new int[1];
        UI.invokeAndWait( new Runnable() {
            public void run() {
                resultHolder[0] = frame.getCursor().getType();
            }
        } );
        return resultHolder[0];
    }
```

Flaky Tests

This test would produce the error about two thirds of the time. In order to get a reliable software failure, we simply repeated the procedure 200 times:

```
boolean noErrorFound = true;
for (int i=0;i<200 && noErrorFound; i++) {
    robot.clickButton( buttonInFrame );
    WaitFor.assertWaitToHaveFocus( buttonInDialog );
    robot.escape();
    noErrorFound = frameCursorType() == Cursor.DEFAULT_CURSOR;
}
TestSetup.disposeAllFrames();
return noErrorFound;
```

How did we come to pick 200 as an appropriate number of repeats? Simply by experiment. With 20 repeats, the test would fail to replicate the bug about one time in ten. It always replicated the bug with 200 repeats. Since the test still took under a minute to execute, it wasn't worth our time to find a smaller number of repeats to use. Once we replaced the JavaWorld code with the code from the Java Specialists' newsletter, the test always passed.

The lesson for us here can be summarized as:

Extreme Testing Guideline: If we pay attention to issues such as thread-safety, it is always possible to write tests that reliably reproduce a bug, and it is vital to do this before we attempt a fix.

Dealing with Flaky Tests

When a test fails, but we're pretty sure that the software under test is working fine, we need to fix the test.

When there is no obvious fix, we should usually proceed as follows. First of all, we modify the test so that when it fails it pauses the test thread and alerts us to the error condition. This will give us the opportunity to inspect the system and identify the problem.

The next step is to repeat the test failure. In the easiest cases, this just means repeatedly running the test in isolation. Some test flakes are more troublesome than this though, and will only occur when dozens of tests have preceded the flaky one. Such was the case with the `IkonMaker` unit test described previously in this chapter.

An efficient technique for identifying the range of tests that need to be run to reliably reproduce the failure works like this. Suppose that we have packages a, b, c, . . . z and a test in package z fails, but not when testing is restricted to z itself. We try running half the tests, say those from n to z. Do we get the failure? If not, we try running half more of the remaining tests, so all tests from g to z. Do we get the failure? If so, we now know that the initial failure point is somewhere between g and n, so we start our testing from halfway between these packages. This **'halving the interval'** technique (the name comes from a proof technique in mathematics) allows us to quickly home in on the test package which, in combination with z, produces the test failure.

For these test flakes the previous test can be a clue as to the cause of the flake. Is there some resource common to both tests? Did the previous test clean up correctly?

Once we have a means of reliably reproducing the test failure, and pausing the broken system, we can start investigating the underlying cause.

Diagnostic Tools

Unfortunately, tools like debuggers and profilers tend to interfere with tests and hence are of not much use in fixing flakes. This is because flakes are often caused by thread race conditions, and the way these pan out differs between runs in a debugger and runs in a normal JVM.

As we have seen, flakes can be caused by test interactions with garbage collection and other resource cleanup tasks. However, the timing of these tasks will generally be altered when running our tests in debuggers, often masking the flakes.

Debuggers tend to slow tests down a lot, and it might simply not be feasible to run the tests in them.

Furthermore, a test involving user interface activity is likely to be severely altered by the existence of the debugger user interface. A remote debugger might be of use in this regard, but is likely to be even slower than a local one.

For these reasons, the best tools to "de-flake" our user interface tests are our own eyes, and System.out is best for other tests.

Tests That Do Not Terminate

Sometimes our test suite will pass, but the JVM in which it ran does not exit at the end of the tests. This is a problem because it prevents Ant or CruiseControl from continuing. It also indicates that there is some resource that has not been cleaned up, which may be a test error or a bug in the production code.

Flaky Tests

Non-terminating tests can be identified using a 'halving the interval' technique similar to that described earlier. Suppose that we have packages a, b, c, . . . z as before, and that when we run all the tests with GTA, the JVM does not terminate. We can try running just the tests a, b, c, . . . l. If these terminate, then our problem, or problems, are in the range m, n, . . . z. This process will pretty soon give us a package of tests that does not terminate. We can then use the same technique within this package to identify the particular class that is causing problems, and once this is done, get to the particular method.

Once a non-terminating method has been found, we start commenting out code to produce an absolutely minimal code snippet that does not terminate. At this point, it should be clear what the problem is. Some common problems are listed below.

Non-Daemon Threads

The JVM will not exit until all non-daemon threads have stopped. If we think that this may be the cause of non-terminating tests, we can press *Control+Scroll-lock* at the command prompt. This will print out a lot of JVM information, including the active threads. If we've named our threads, this will tell us which ones are still alive.

Remote Objects

Remote objects that were exported but have not been unexported will keep a JVM alive. Here is the `cleanup()` method from one of our tests that had this problem. The fields `server` and `adminCM` are remote objects that we forcibly un-export. The code is pretty rough, with lots of empty catch clauses, but it does the job well enough.

```java
private void cleanup() {
    try {
        server.shutdown("");
    } catch (Exception e) {
    }
    try {
        unexport(server);
        unexport(adminCM);
    } catch (Exception e) {
    }
    try {
        Naming.unbind(SERVICE_NAME);
    } catch (Exception e) {
    }
    TestSetup.disposeAllFrames();
    TestSetup.gc();
}
```

```
    private void unexport(Remote remote) {
        try {
            UnicastRemoteObject.unexportObject(remote, true);
        } catch (Exception e) {
        }
    }
}
```

Server Socket Still Waiting

One way a non-daemon thread can be prevented from terminating is by having it listen on a server socket. Here's a program that demonstrates this:

```
    public static void main( String[] args ) throws IOException {
        Thread thread = new Thread() {
            public void run() {
                try {
                    ServerSocket server = new ServerSocket( 8888 );
                    server.accept();
                } catch (IOException e) {
                }
            }
        };
//          thread.setDaemon( true );//Program terminates.
        thread.setDaemon( false );//Program does not terminate.
        thread.start();
    }
```

Frame Not Properly Disposed

Frames that are not disposed of properly keep the AWT event queue, and hence the JVM, alive, as this little program demonstrates:

```
    public static void main( String[] args ) {
        JFrame frame = new JFrame( "Junk" );
        UI.showFrame( frame );
        frame.setVisible( false );
//        UI.disposeOfAllFrames();
    }
```

Un-commenting the fifth line allows the program to terminate.

Summary

Flaky tests and tests that do not terminate undermine our confidence in our software. Further, they eat into our productivity by preventing a continuous integration process. For these reasons, we need to eliminate them. In this chapter, we've seen ways of helping to track them down, which is the first step to fixing them. We've also seen some of the common causes of non-terminating tests.

Index

A

acceptance testing 30
application level testing 30
automated test execution
  about 31
  justifications 32

C

class serialization
  testing 209
  unit test 210
client-side
  adduser class 215
  adduser class, implementing 216
  implementing 215
  user, creating 214
Cobertura tool 28
combo boxes
  testing 160, 161
comment editor
  about 82
  testing, interfaces used 90
comment manager
  testing 82
component
  Frame,allFrames() method 110
  getComponents() method 110
  searching 110, 111
  searching, by name 113, 114
  state, reading 114
concurrent modifiers
  about 179, 181
  testing 179
concurrent readers
  testing 181, 182

concurrent writers
  testing 181, 182
console-based programs output
  output, reading from a separate
      JVM 196-198
  stream-switching, testing strategies
      194, 195
  testing 193
  testing strategies 193
context-sensitive help
  creating 172-174
  testing 172-175
Cyborg
  about 59
  basic keyboard operations 60-62
  basic mouse operations 63
  design 59
  features 59
  keyboard data, entering 60
  mouse, dragging 64
  mouse operations 63
  screen, checking 65

D

database
  accessing 224
  database management 229, 230
  databasemanager class, defining 224
  low level database transactions,
      wrapping 224
  manager class 226
  persistence testing 228, 229
  persistence testing, steps 228
  storable class 226
  user class 227

Dbunit tool 223
DeleteCase test
   helper classes 237
   implementing 237-239
Design by Contract
   iContract tool 42
   post-conditions 41
   post-conditions, benefits 41
   pre-conditions 40
diagnostic tools
   about 297
   debuggers 297
   profilers 297
distributed testing
   example 286, 287
   GTA used 285
   working 285, 286

E

email
   testing 201
   testing, external email account used 202-204
   testing, local email server used 205, 206
   testing with remote account, difficulties 204
EMMA tool 27
Excel spreadsheets
   testing 206, 207
explicit unit test 17
extreme testing
   about 19
   guideline 19

F

file management
   operations, in testing 212
   testing 211
flaky tests
   about 289
   dealing with 296
   Ikon Do It unit test 289-294
frames
   frame location, testing 147, 148
   frame size, testing 149
   Ikonmaker unit test, for frame location 147
   key points, of IkonMaker unit test 148

   testing 147
functional testing 30
function tests
   implementing, by developers 31
   JVMs with GUI components 242
   JVM using 241, 242
   LabWizard example 234
   sever, testing in its own JVM 240
   specification 233
   using, as tutorial 247-249
   Validator UI Wrapper class 243
   web service, testing 251-255

G

GroundTestAuto. *See* GTA
GTA
   about 262, 269
   advantages 273
   classes, need no testing 280
   distributed testing 285
   distributed testing, example 286, 287
   distributed testing, working 285, 286
   extra unit tests 280
   LabWizard product suite 270
   overload methods, testing 277, 278
   parameters 284, 285
   parameters file 271
   protected methods, testing 279
   unit test coverage 272
   unit test coverage, working 274-276
   unitTester class 271
   uses 281
GTA, uses
   GTA, running from Ant or Cruisecontrol 284
   individual test methods, running 283
   one level test, running 281
   package name abbreviation 283
   tests for selection of classes in single package, running 283
   tests for selection of packages, running 282
   unit tests for single class, running 282
   unit tests for single package, running 281
GUI applications
   testing 16

H

HelpGenerator
  executing 175
hidden constructors
  invoking 25-27
hidden methods
  invoking 25-27
hierarchy of test
  function tests 33
  load tests 33
  unit tests 33

I

iContract tool 42
Ikon Do It example
  IkonCanvas, testing 144-146
  IkonCanvas unit test 145
  Save as dialog 119
  Save as dialog, testing 120, 121

J

JavaHelp package
  overview 168, 169
JavaHelp system
  broken links, testing 171
  context-sensitive help, creating 172
  context-sensitive help, testing 172
  file content, testing 169
  HelpGenerator, executing 175
  help pages, testing 171
  HTML file, testing 170
  indexes, testing 169
  index item, testing 171
  structural level, testing 169
  tests, need for 169
JColorChooser
  testing 139-141
Jcomponent appearance
  testing 144
JFileChooser
  about 142
  properties 142
  properties, testing 143
  setup 142

JMenus
  items, testing 156-158
  menus, using with Cyborg 159
  testing 156
JPopupMenus
  testing 160
JProgressBar
  testing 161
JSlider
  testing 163, 164
JSpinner
  testing 163, 164
JTable
  testing 153-155
  unit test 154
JTree
  testing 164, 165
JVM
  active threads 183
  active threads, example 184-186
  active threads, testing 183
  threads, counting 189

L

LabWizard example
  LabWizard comment editor 81, 82
  threads 177
LabWizard example, function tests
  administrator 234
  documentation and code hierarchy 234
  knowledge builder 234
  Rippledown Product Spec 234
  server 234
  System Design Spec 235
  validator 234
LabWizard example, load tests
  BMD server, issues 260
  housekeeping, issues 259
  issues 258
  rejected cases, issues 259
LabWizard login screen
  data verification 93
  demo page 91
  state properties, checking 92
  testing 92
  usability criteria 93

legacy code
  testing 29
Liskoy Substitution Principle 96
lists
  list, rendering 151, 152
  list properties 153
  list selection methods 150
  testing 150
load testing 49
load tests
  Apache JMeter used 267
  for LabWizard 265
  GTA specification 261
  GTA structure 262
  GTA test 262, 263
  jet.testtools.Size class 264
  profilers 266
  RAM usage, measuring 264
  RAM usage, measuring size class used 265
  stopwatch class 261
  time, measuring 261
  time, measuring stopwatch class used 260
  writing 258
log files
  cleaning up 193
  testing 191, 192
LoginScreen
  about 94
  code outline 94, 95
  design 94
  helper class 94
  implicit tests 105
  LoginScreen class 94
  LoginScreen.Handler 99
  testing 102-104
  UILoginScreen 97
  UI wrappers 96
  unit test, setting up 100, 101
LoginScreen.Handler
  implementing 100

M

multi-threading systems, testing techniques
  concurrent modifiers 179
  concurrent readers 181
  concurrent writers 181
  JVM threads, testing 183
  proof of thread completion 183
  threads, couting 189
  waitForNamedThreadToFinish(),
    testing 187
  waiting class 177
MVC system
  about 214, 217
  issues 217
  server-side class 214
  user class, implementing 214

N

non-terminating tests
  about 297
  frame, not disposed 299
  non-daemon threads 298
  remote objects 298
  server socket 299

P

PDF documents
  testing 208, 209
private classes
  methods, need no testing 24, 25
  need no testing 21, 22
profilers 266
progress bars
  testing 161
public classes
  methods, need for testing 22-24
  need for testing 16

R

reliable tests
  writing 294-296
resource bundles
  about 67
  issues 67, 68
  single class approach 69
  single class approach, advantages 69
  single class approach, working 69
  solutions, for issues 69
  UserStrings class 70

ResourcesTester
  about 73
  working 74-78
Robot 59

S

server-side class
  advantages 221
  issues 217
  solution 218-220
softwares
  log files, testing 191
step method
  testing 86-89
Swing components
  invokeAndwait(Runnable r) method 108
  invokeLater(Runnable r) method 108
  javax.swing.swingUtilities class 108
  manipulating 108
  setting up, in thread-safe manner 108, 109
swing threading 117
system testing
  about 30
  sub-categories 30

T

test data
  managing 54-57
test data management
  prerequisites 55
test first approach 19
test infrastructure
  issues 51
test infrastructure, settingup
  funcion tests, packaging 53
  load tests, packaging 54
  temporary files 57
  test data management 54
  test packages, organizing 52
  unit tests, packaging 51, 52
testing
  about 13
  application view 13
  at application level 30
  case study 115
  Cyborg class 59

module view 13
necessity 19, 20
platform 34
unit view 13
views 13

U

UI helper methods
  dialogs 123
  frame disposal 125
  getText() method 124
UILoginScreen
  implementing 98
UI wrappers
  about 96
  issues 97
  public methods 97
  public methods, implementing 97
  UILoginScreen, implementing 98
unit test
  action can be cancelled, testing for 115, 116
  bootstrapping 48
  constructor test 130, 131
  data validation test 135, 136
  default implementation 38
  Design by Contract 40
  documentation 29
  example 37
  features 18
  Ikon Do It example 119
  implementing, by developers 28
  infrastructure 125
  integration 29
  Lab Wizard example 14, 16
  name() test 133
  show() test 134
  test cases 39, 40
  test code example 45, 47
  usability test 136
  wasCancelled() test 131, 132
unit test, Ikon Do It example
  outline 121, 122
unit test, LoginScreen
  cleanup() method 101
  error messages, testing for 112
  init() method 101

   setting up 100
   show() method 101
unit test coverage 27
unit test infrastructure
   cleanup() method 129
   init() method 128
   ShowerThread class 127
   UISaveAsDialog class 125, 126
user interface components
   setting up, in thread-safe manner 108
UserStrings class
   about 70, 71
   JButton method 78
   JMenuItem method 78
   resource bundle 72
   subclass 72

W

waitForNamedThreadToFinish()
   unit test 187, 188
waiting class
   about 177, 178
   testing 177
wizard
   about 83
   step method 84
   testing 83, 85

Packt Open Source Project Royalties

When we sell a book written on an Open Source project, we pay a royalty directly to that project. Therefore by purchasing Swing Extreme Testing, Packt will have given some of the money received to the GrandTestAuto project.

In the long term, we see ourselves and you—customers and readers of our books—as part of the Open Source ecosystem, providing sustainable revenue for the projects we publish on. Our aim at Packt is to establish publishing royalties as an essential part of the service and support a business model that sustains Open Source.

If you're working with an Open Source project that you would like us to publish on, and subsequently pay royalties to, please get in touch with us.

Writing for Packt

We welcome all inquiries from people who are interested in authoring. Book proposals should be sent to authors@packtpub.com. If your book idea is still at an early stage and you would like to discuss it first before writing a formal book proposal, contact us; one of our commissioning editors will get in touch with you.

We're not just looking for published authors; if you have strong technical skills but no writing experience, our experienced editors can help you develop a writing career, or simply get some additional reward for your expertise.

About Packt Publishing

Packt, pronounced 'packed', published its first book "Mastering phpMyAdmin for Effective MySQL Management" in April 2004 and subsequently continued to specialize in publishing highly focused books on specific technologies and solutions.

Our books and publications share the experiences of your fellow IT professionals in adapting and customizing today's systems, applications, and frameworks. Our solution-based books give you the knowledge and power to customize the software and technologies you're using to get the job done. Packt books are more specific and less general than the IT books you have seen in the past. Our unique business model allows us to bring you more focused information, giving you more of what you need to know, and less of what you don't.

Packt is a modern, yet unique publishing company, which focuses on producing quality, cutting-edge books for communities of developers, administrators, and newbies alike. For more information, please visit our website: www.PacktPub.com.

Java EE 5 Development using GlassFish Application Server

ISBN: 978-1-847192-60-8   Paperback: 400 pages

The complete guide to installing and configuring the GlassFish Application Server and developing Java EE 5 applications to be deployed to this server

1. Concise guide covering all major aspects of Java EE 5 development
2. Uses the enterprise open-source GlassFish application server
3. Explains GlassFish installation and configuration
4. Covers all major Java EE 5 APIs

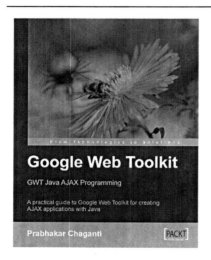

Google Web Toolkit GWT Java AJAX Programming

ISBN: 978-1-847191-00-7   Paperback: 240 pages

A step-by-step to Google Web Toolkit for creating Ajax applications fast

1. Create rich Ajax applications in the style of Gmail, Google Maps, and Google Calendar
2. Interface with Web APIs create GWT applications that consume web services
3. Completely practical with hands on examples and complete tutorials right from the first chapter

Please check www.PacktPub.com for information on our titles

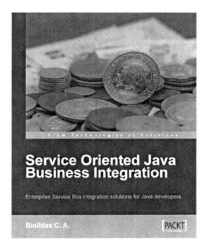

Service Oriented Java Business Integration

ISBN: 978-1-847194-40-4　　　Paperback: 414 pages

Enterprise Service Bus integration solutions for Java developers

1. Vendor-independent integration of components and services through JBI explained with real-world examples
2. Hands-on guidance to ESB-based Integration of loosely coupled, pluggable services
3. Enterprise Integration Patterns (EIP) in action, in code
4. ESB integration solutions using Apache open-source tools

Quality Assurance for Dynamics AX-Based ERP Solutions

ISBN: 978-1-847192-91-2　　　Paperback: 168 pages

Verifying Dynamics AX customization to the Microsoft IBI Standards

1. Learn rapidly how to test Dynamics AX applications
2. Verify Industry Builder Initiative (IBI) compliance of your ERP software
3. Readymade testing templates
4. Code, design, and test a quality Dynamics AX-based ERP solution

Please check www.PacktPub.com for information on our titles

Printed in the United States
113657LV00005B/49-50/P